Modern Thermoelectrics

Modern Thermoelectrics

D. M. Rowe MSc, PhD

Lecturer in Applied Physics
University of Wales Institute of Science and Technology

AND

C. M. Bhandari MSc, PhD

Lecturer in Physics
University of Allahabad, India

HOLT, RINEHART AND WINSTON
LONDON · NEW YORK · SYDNEY · TORONTO

Holt, Rinehart and Winston Ltd: 1 St Anne's Road,
Eastbourne, East Sussex BN21 3UN

British Library Cataloguing in Publication Data

Rowe, D. M.
Modern thermoelectrics.
1. Thermoelectricity
I. Title II. Bhandari, C. M.
621.31'43 QC621

ISBN 0–03–910433–8

Photoset and printed in Malta by Interprint Limited

Last digit is print number: 9 8 7 6 5 4 3 2 1

Preface

I dive down into the depths of the ocean of forms
hoping to gain the perfect pearl of the formless

Tagore

THE thermoelectric effects which underlie the conversion of heat energy into electrical energy and vice versa have been known for more than 150 years but it is only during the last 20 years that the development of semi-conductor thermoelements has made possible the fabrication of practical thermoelectric generators and refrigerators. The use of thermoelectric devices has to date been restricted to specialized applications where combinations of their desirable properties outweigh their relatively low efficiency and high cost. Recently, however, there have been significant advances in material technology and the indications are that in the near future thermoelectric materials with substantially improved performance will become available and this may result in a more widespread application of thermoelectric devices.

This book is intended to serve both as an introduction to thermoelectric generation and refrigeration and as an updated source of information on thermoelectric material development and device applications. The first chapter gives a brief history of thermoelectricity. This is followed by a basic chapter in which the principal thermoelectric coefficients and parameters associated with thermoelectric energy conversion are defined. Chapters 3 and 4 are mainly mathematical. Chapter 3 concentrates on elementary solid state theory and the thermoelectric transport equations, while Chapter 4 deals with the optimization of the thermoelectric figure of merit, and both highly idealized and realistic models of thermoelectric materials are used in examining its theoretical upper limit. Established thermoelectric materials and recent advances in material technology are discussed in Chapter 5 and the various methods used in measuring the parameters which occur in the thermoelectric figure of merit described. Chapter 6 deals with thermocouple construction and configuration, while Chapters 7 and 8 cover thermoelectric refrigerators and generators respectively. Chapter 9 is devoted entirely to applications of thermoelectric generators. These are divided into two types, non-isotope and isotope powered. Emphasis is placed on the latter with their applications and environmental consequences discussed in some detail. Finally Chapter 10 reviews recent progress in thermoelectric energy conversion and assesses its future prospects.

The tireless efforts of the Technology Editor, Mr G. Newham, and the Manuscript Editor, Mr G. Russel, at all stages of the book's preparation are greatly appreciated and we would like to thank Professor H. J. Goldsmid and Dr J. D. Lewins for reading the manuscript and making a number of helpful suggestions.

To Our Parents

Front cover
Astronaut Bean cautiously removes the hot fuel capsule from its graphite cask in order to insert it into the Radioisotope Thermoelectric Generator (RTG) in the foreground. The temperature of the capsule which was filled with plutonium-238 is about 1350 K. The SNAP-27 RTG was the powerhouse of the entire lunar experimental package. The temperature difference between the fuel capsule and the finned outer housing was converted into electrical power by 442 lead telluride thermocouples giving about 74 watts of electrical power (Courtesy National Aeronautics and Space Administration).

Contents

1

An Introduction

In 1821 Seebeck[1] reported the results of experiments in which a compass needle was deflected if placed in the vicinity of a closed loop, formed from two dissimilar conductors, when one of the junctions was heated. Seebeck erroneously concluded that the interaction was a magnetic phenomenon and, in pursuing this line of thought, attempted to relate the earth's magnetism to the temperature difference between the equator and the poles. Nevertheless, he did investigate the phenomenon in a large number of materials, including some we now call semiconductors, and arranged them in order of magnitude of the product $\alpha\sigma$, where α is the Seebeck coefficient and σ the electrical conductivity. The Seebeck coefficient is expressed in volts per degree or, more often, μV/K. The 'Seebeck Series' formed in this way is very similar to the present-day thermoelectric series and, had Seebeck employed the first and last members of his series in a thermocouple, he could have converted thermal energy into electricity in 1821 with an efficiency of $\sim$3 per cent, which compares very favourably with the most efficient steam engine of the day. With the benefit of hindsight it is apparent from Seebeck's account that the phenomenon observed was caused by an electric current flowing in the circuit, and that he had discovered the first of the so-called thermoelectric effects.

Some twelve years later, a complementary effect was discovered by Peltier[2] who observed temperature changes in the vicinity of the junction between dissimilar conductors when a current passed. Although Peltier used the Seebeck effect in his experiments as a source of weak currents, he failed to appreciate the fundamental nature of his observations, or to relate the Peltier effect to the findings of Seebeck. The true nature of the Peltier effect was explained by Lenz[3] in 1838. He concluded that, depending on the direction of current flow, heat is absorbed or generated at a junction between two conductors, and demonstrated this by freezing water at a bismuth-junction and melting the ice by reversing the direction of current flow.

The lack of interest and slow progress in thermoelectric application which followed the discovery of thermoelectric phenomena is understandable when one recalls that much more exciting discoveries were made during this period. This was the era of electromagnetism, with the initial discoveries of Oersted being followed by investigation of researchers such as Ampère and Laplace, and culminating in the laws of electromagnetic induction by Faraday.

Thermoelectricity enjoyed a temporary revival from 1850, with the development of thermodynamics, when interest focused on all forms of energy conversion. In 1851[4] W. Thomson (Lord Kelvin) established a relationship between the Seebeck and Peltier coefficients and predicted the existence of a third thermoelectric effect, the Thomson effect, which he subsequently observed experimentally. This effect relates to the heating or cooling in a single homogeneous conductor when a current passes along it in the presence of a temperature gradient.

The possibility of using thermoelectric phenomena in the generation of electricity was considered in 1885 by Rayleigh who first calculated, although incorrectly, the efficiency of a thermoelectric generator. In 1909[5] and 1911[6] Altenkirch gave a satisfactory theory of thermoelectric generation and refrigeration and showed that good thermoelectric materials should possess large Seebeck coefficients with low thermal conductivity, to retain the heat at the junction, and low electrical resistance to minimize Joule heating. These desirable properties were embodied in a so-called figure of merit Z, where $Z=\alpha^2\sigma/\lambda$. λ is the thermal conductivity and the unit of Z is 1/K. At a given absolute temperature T, since Z may vary with T, a useful non-dimensional figure of merit will be ZT.

Although the properties favourable for thermoelectric applications were well known, the important advantages offered by Seebeck's mineral semiconductors were overlooked with the attention of researchers remaining focused on metals and metal alloys. In these materials the ratio of thermal conductivity to electrical conductivity is a constant (Wiedemann–Franz–Lorenz law) and it is not possible to reduce one while increasing the other. Consequently, the metals best suited are those with the highest Seebeck coefficients. Most metals possess Seebeck coefficients of 10 μV/K or less, giving associated generating efficiencies of a fraction of 1 per cent, which is uneconomical as a source of electrical power. Similar considerations also led to the conclusion that thermoelectric refrigeration was an uneconomical proposition.

Renewed interest in thermoelectricity accompanied the development in the late 1930s of synthetic semiconductors that possessed Seebeck coefficients in excess of 100 μV/K, and in 1947 Telkes[7] constructed a generator that operated with an efficiency of about 5 per cent. In 1949 Ioffe[8] developed a theory of semiconductor thermoelements. A prototype domestic refrigerator was constructed in 1953 which had a coefficient of performance of around 0.2 when the temperature inside the refrigerator was 24 K below room temperature.

Unfortunately, in semiconductors the ratio of the thermal to electrical conductivity is greater than in metals owing to their poorer electrical conductivity. It was not obvious that semiconductors were superior thermoelectric materials and, apart from Soviet activities, interest once again waned. Research into compound semiconductors for possible transistor application in the 1950s resulted in new materials with substantially improved thermoelectric properties, and in 1956 Ioffe and his co-workers[9] demonstrated that the ratio λ/σ could be decreased if the thermoelectric material is alloyed with an isomorphous element or compound. Spurred on by possible military applications a tremendous survey of materials was undertaken, particularly at the RCA laboratories in the USA, which resulted in the discovery of a few semiconductors with ZT values approaching 1.5.

In the early 1960s a requirement for autonomous sources of electrical power arose from the exploration of space, advances in medical physics and the exploitation of the earth's resources in increasingly hostile and inaccessible locations. Thermoelectric generators are ideally suited to such applications: they are compact, have no moving parts, and are silent in operation. Advantage can be taken of the simplicity and ruggedness of thermoelectric generators compared with thermal–mechanical conversion devices.[10] In situations where periodic refuelling is possible and oxygen available, fossil fuel is employed as a heat source. Hydrocarbon fuel has an energy density some 50 times that of a chemical battery and so, provided that the conversion efficiency is better than 2 per cent, a hydrocarbon-fuelled system can provide a much lighter and less bulky source of long-term electrical energy than batteries. When annual refuelling is not possible, or oxygen is not available, radioactive isotopes serve as heat sources, enabling the generators—which are referred to as radioisotope thermoelectric generators or RTGs—to operate unattended for extended periods, in some instances for longer than 15 years.

Over the last decade thermoelectric generators have found use in specialized applications in which combinations of their desirable features outweigh their relatively high cost and low generating efficiencies, typically ~5 per cent. A thermoelectric generator in essence consists of an arrangement of thermocouples, referred to as a module. The efficiency of conversion of heat into electricity depends on the temperature difference, ΔT, over which the device operates, on its average temperature of operation, $\bar{T}$, and on the performance of the thermocouple material through its figure of merit, Z. In recent years, emphasis has been placed on developing high-temperature semiconductor thermocouple materials such as Si–Ge alloys, which possess relatively stable properties up to 1000 K and permit the utilization of high grade heat sources, such as chemical flame and isotopic radiations. Boron–carbon compounds are being developed for use at substantially higher temperatures than silicon–germanium alloys.[11,12]

Thermoelectric refrigeration has also enjoyed some success in domestic food refrigerators, air conditioning and in numerous novel applications where the facility to vary the cooling capacity of the device to match the particular application has proved to be an important factor.

Following the fivefold increase in the price of crude oil in 1974, a closer look has been taken at the possibility of large scale production of electricity by the thermoelectric effect. At present the maximum module conversion efficiency claimed with a two stage generator is around 13.5 per cent[13] and clearly does not compete with a central power station efficiency of approximately 40 per cent. Apart from an abundant supply of easily utilized heat, large scale thermoelectric generation of electricity will require the cheap production of substantial amounts of semiconductor material, accompanied by a significant improvement in the material figure of merit. While this indicates that large scale application of thermoelectricity is not imminent, the specialized applications are far more promising and have motivated a search for materials with an improved figure of merit, Z.

The development of improved thermoelectric material is in the first instance closely dependent upon the formulation of an adequate theoretical model. Solid state theory has greatly assisted in this direction although the models available are, at best, rough approximations to the actual materials. In recent years the upper limit to the material figure of merit has been investigated, but realization of the same in practice depends upon many factors, not the least of which is whether or not the material can be prepared.

Although advances are being made in thermoelectric refrigeration technology it is in thermoelectric generation that the vast majority of research effort is being concentrated. Superior high temperature thermoelectric materials are currently under development for use in radioisotope thermoelectric generators (RTGs) to provide onboard power to spacecraft assigned to deep space missions. The successful development of materials with improved thermoelectric properties is difficult. In 1973 the selenide isotope generation (SIG) programme began; its objective was the production of high performance generators for use on the scheduled 1985 Galileo mission.[14] Unfortunately the programme was terminated in 1979 due to difficulties associated with the production of stable selenide thermoelectrics. Subsequently generators based upon the well-established silicon-germanium technology were selected for the Galileo mission. Attempts to improve the performance of materials based upon the silicon-germanium alloys continue with the major effort directed at reducing the lattice thermal conductivity by introducing additional disorder into the alloy structure. A recent analysis by the Fairchild Space and Electronics Company of a multi hundred watt system using silicon–germanium–gallium phosphide thermoelements indicates that a system efficiency of 9.4 per cent should be achieved by the late 1980s.[14] The preparation of the materials involves advanced and expensive technology which is not readily adaptable to large-scale production. In applications where near approaches to the sun are envisaged, solar thermoelectric generators (STGs) are being considered.[15] Hot pressing methods of preparation hold out the possibility of cheap production of established materials on a large scale and offer as a bonus possible improvements in the material figure of merit.[16,17]

Further research into and development of thermoelectrics is assured with the continuation of the US space programmes, and this at a level of funding that almost ensures a significant breakthrough in thermoelectric generation. Advances in other applications are likely to be less dramatic and derived as a spin off from National Aeronautics and Space Administration (NASA) activities.

REFERENCES

1. T.J. Seebeck, 'Magnetische Polarisation der Metalle und Erze durch Temperatur-Differenz', *Abhandlungen der Deutschen Akademie der Wissenschaften zu Berlin*, (1822), 265–373.
2. J.C. Peltier, 'Nouvelles expériences sur la caloricité des courans electriques', *Annales de chimie*, **LVI** (1834), 371–87.
3. See A.F. Ioffe, *Semiconductor Thermoelements and Thermoelectric Cooling*, Infosearch, London, 1957.
4. W. Thomson, 'On a mechanical theory of thermoelectric currents', *Proceedings of the Royal Society of Edinburgh*, (1851), 91–8.
5. E. Altenkirch, 'Uber den Nutzeffekt der Thermosaule', *Physikalische Zeitschrift*, **10** (1909), 560–8.
6. E. Altenkirch, 'Electrothermische Kalteerzeugung und Reversible Electrische Heizung', *Physikalische Zeitschrift*, **12** (1911), 920–4.
7. M. Telkes, 'The efficiency of thermoelectric generators', *International Journal of Applied Physics*, **18** (1947), 1116–27.
8. A.F. Ioffe, *Energeticheskic osnovy termoelektricheskikh baterei iz poluprovoduikov*, Academy of Sciences of the USSR, Moscow, 1949.
9. A.F. Ioffe, see *Dokladȳ Akademie nauk SSSR*, **106** (1956), 981.
10. E.H. Cooke-Yarborough et al., *The Harwell Thermomechanical Generator*, UKAEA publication AERE-R7714, 1974.

11. C. Woods, 'High temperature thermoelectric properties of boron–carbon compounds', *Proceedings of Fourth International Conference on Thermoelectric Energy Conversion*, (1982), University of Texas at Arlington, panel session.
12. G.H. Reynolds, N.B. Elsner and J.H. Norman, 'Thermoelectric properties of ternary boron–carbon based compositions', *Proceedings of Fourth International Conference on Thermoelectric Energy Conversion*, (1982), University of Texas at Arlington, paper II (5).
13. D. Puotinen, see F.D. Rosi, 'Thermoelectricity and thermoelectric power generation', *Solid State Electronics*, **11** (1968), 833–68.
14. P.A. O'Riordan, 'The US Department of Energy's thermoelectric development for space power', *Proceedings of Fourth International Conference on Thermoelectric Energy Conversion*, (1982), University of Texas at Arlington, 15–25.
15. V. Raag and L. Hankins, 'Design concepts of solar thermoelectric generators in space applications', *Proceedings of Second International Conference on Thermoelectric Energy Conversion*, (1978), University of Texas at Arlington, 60–5.
16. C.M. Bhandari and D.M. Rowe, 'Silicon–germanium alloys as high temperature thermoelectric materials', *Contemporary Physics*, **21** (3) (1980), 219–42.
17. D.M. Rowe, 'The development of fine grained silicon–germanium alloy technology at UWIST', *Proceedings of Fourth International Conference on Thermoelectric Energy Conversion*, (1982), University of Texas at Arlington, paper IV (2).

2

Definitions and Basic Principles

THE basic concepts of thermoelectricity have been described in a number of books[1-6] and review papers[7-12] and need only be dealt with briefly here.

2.1 THERMOELECTRIC COEFFICIENTS

The Seebeck effect is the phenomenon underlying the conversion of heat energy into electrical power. Its physical significance can be appreciated by considering the effect of imposing a steady temperature gradient along a finite conductor. Initially the conductor possesses a uniform distribution of charge carriers, but in the presence of a temperature gradient the free carriers at the hot end will have a greater kinetic energy and tend to diffuse to the cold end. The build up of charge results in a back emf which opposes a further flow of charge. The open circuit voltage when no current flows is the Seebeck voltage.

If the junctions of a circuit formed from two dissimilar conductors a and b (Figure 2.1 position 1), which are connected electrically in series but thermally in parallel, are maintained at different temperatures T_1 and T_2 and $T_1 > T_2$, an open circuit potential difference, V, is developed and is given by $V = \alpha_{ab}(T_1 - T_2)$. For small temperature differences the relationship is linear and defines the Seebeck coefficient α_{ab} for the junction. The Seebeck coefficient is measured in V/K or more often in μV/K. V and α are considered positive if the conventional current tends to flow from a to b at the hot junction. Thus the sign of the Seebeck coefficient as well as its magnitude will depend upon the choice of materials.

The complementary effect, the Peltier effect (Figure 2.2), is the phenomenon used in thermoelectric refrigeration, with the rate of reversible heat absorption $\dot{Q}$ accompanying the passage of a current I through the junction given by $\dot{Q} = \pi_{ab} I$. π_{ab} is the Peltier coefficient of the junction and is measured in watts per ampere or in volts. By convention, π_{ab} is taken as positive when current flows from a to b and Q is absorbed from the surroundings. This heat may also include Joule heat from the resistance of the junction. The Peltier effect arises because the potential energies of the charge carriers in the materials on either side of the junction are different, as are the scattering mechanisms which govern the equilibrium between the charge carriers and the crystal lattice.

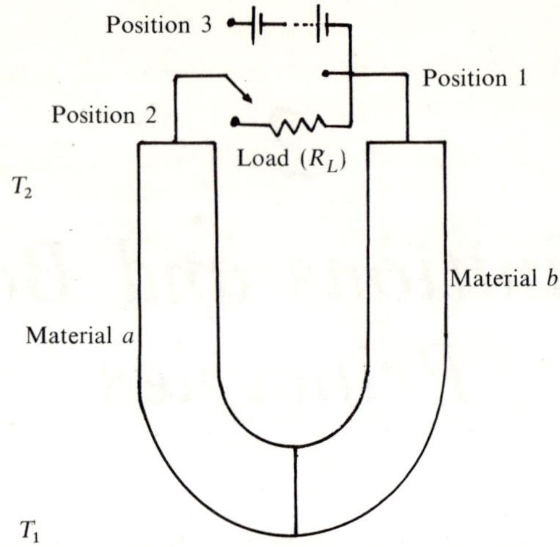

Figure 2.1 Schematic thermoelectric device.

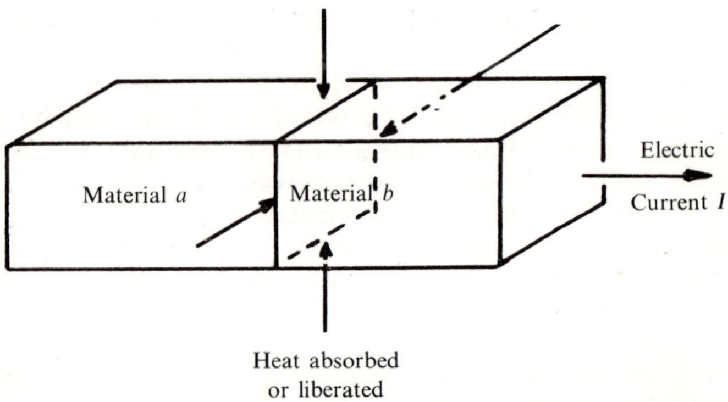

Figure 2.2 Peltier effect.

Consequently, when a current flows through the junction, energy must be interchanged with the surroundings in order to maintain conservation of energy and charge.

The last of the thermoelectric effects, the Thomson effect (Figure 2.3), relates to $\dot{Q}$ the rate of generation of reversible heat, which results from the passage of a current along a portion of a single conductor over which there is a temperature difference ΔT. Providing the temperature difference is small, $\dot{Q}=\beta I \Delta T$ where β is the Thomson coefficient. The units of β are the same as those of the Seebeck coefficient: V/K. As the heat flow depends upon the product of I and ΔT, the Thomson effect is not of primary importance in thermoelectric devices but should not be neglected in detailed calculations. The origin of the effect is essentially the same as the Peltier effect but with the temperature gradient along the conductor now responsible for differences in potential energy of the charge carriers and in the scattering mechanisms.

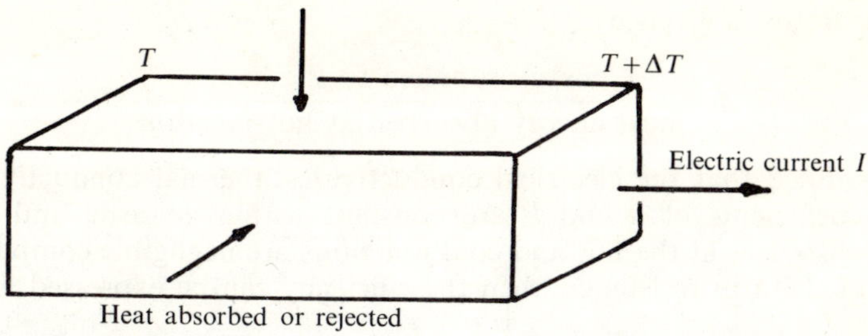

Figure 2.3 Thomson effect.

The three thermoelectric coefficients are related by the kelvin relationships:

$$\alpha_{ab}=\frac{\pi_{ab}}{T} \quad \text{and} \quad \frac{d\alpha_{ab}}{dT}=\frac{\beta_a-\beta_b}{T} \qquad 2.1$$

These relationships can be derived using the Onsager reciprocal relationships and the standard theory of irreversible thermodynamics.[13] The validity of these relationships has been demonstrated for a number of thermocouple materials[14] and it is assumed that they hold for all materials used in thermoelectric application.

Both the Seebeck and Peltier coefficients are defined in terms of the properties of a junction, whereas the Thomson coefficient is the property of a single conductor. The second relationship enables a definition to be derived for the Seebeck coefficient of a single material and is given by:

$$\alpha=\int_0^T \frac{\beta}{T}\,dT \qquad 2.2$$

The Seebeck coefficient is negative for *n*-type semiconductors and positive for *p*-type semiconductors.

Another important relationship derived from the kelvin relations is that the Seebeck coefficient of the junction between two materials *a* and *b* is the same as the difference between the two absolute coefficients. Thus if the absolute coefficient of one of the junction materials is known, the other can be obtained by measuring the Seebeck coefficient for the junction. Seebeck coefficients are generally quoted with reference to lead. Superconductors possess zero thermoelectric coefficients, and the absolute Seebeck coefficient of lead up to 18 K has been determined from the voltage of a lead superconductor thermocouple, and up to room temperature from indirect measurements of the Thomson heat of an alloy and the Seebeck coefficient of lead relative to the same alloy.[15] More recent measurements of the Seebeck coefficient of lead between 10 K and 350 K by direct measurement of the Thomson heat indicate that at about 20 K presently accepted values may be inaccurate.[16]

2.2 GENERATING EFFICIENCY AND THE FIGURE OF MERIT

When load resistance R_L is connected across the cold ends of thermocouple arms *a* and *b* (Figure 2.1, position 2) heat energy supplied at the hot junction will drive a current through the circuit and deliver power to R_L. The efficiency of the generator is given by:

$$\phi = \frac{\text{energy supplied to the load}}{\text{heat energy absorbed at hot junction}} \tag{2.3}$$

If it is assumed that the electrical conductivities, thermal conductivities and Seebeck coefficients of *a* and *b* are constant within an arm, and that the contact resistances at the hot and cold junctions are negligible compared with the sum of the arm resistance, then the efficiency can be expressed as:

$$\phi = \frac{I^2 R_L}{\alpha_{ab} I T_1 + \lambda'(T_1 - T_2) - \frac{1}{2} I^2 R} \tag{2.4}$$

where λ' is the thermal conductance of *a* and *b* in parallel and *R* is the series resistance of *a* and *b*. In thermoelectric materials σ, λ' and α change with temperature, and in both generation and refrigeration this is taken into account;[17] however, the simple expression obtained for the efficiency can still be employed with an acceptable degree of accuracy if approximate average values are adopted for these parameters over the temperature range of interest.[18] Appropriate allowances can also be made for contact resistance.[19]

Efficiency is clearly a function of the ratio of the load resistance to the sum of the generator arm resistances, and at maximum power output it can be shown[20] that:

$$\phi_p = \frac{T_1 - T_2}{\frac{3T_1}{2} + \frac{T_2}{2} + \frac{4}{Z_c}} \tag{2.5}$$

while the maximum efficiency:

$$\phi_{max} = \eta_c \gamma \tag{2.6}$$

where

$$\eta_c = \frac{T_1 - T_2}{T_1} \tag{2.7}$$

$$\gamma = \frac{\sqrt{1 + Z_c \bar{T}} - 1}{\sqrt{1 + Z_c \bar{T}} + \frac{T_2}{T_1}} \tag{2.8}$$

and

$$\bar{T} = \frac{T_1 + T_2}{2} \tag{2.9}$$

and

$$Z_c \text{ (the figure of merit of the couple)} = \frac{\alpha_{ab}^2}{R\lambda'} \tag{2.10}$$

The maximum efficiency is thus the product of the Carnot efficiency, which is clearly less than unity, and γ, which embodies the parameters of the materials.

If the geometries of *a* and *b* are matched to minimize heat absorption, then:

$$Z_c = \frac{\alpha_{ab}^2}{\left[\left(\frac{\lambda_a}{\sigma_a}\right)^{\frac{1}{2}} + \left(\frac{\lambda_b}{\sigma_b}\right)^{\frac{1}{2}}\right]^2} \tag{2.11}$$

In practice the two arms of the junction have similar material constants, in which case the concept of a figure of merit for a material is employed and given by:

$$Z = \frac{\alpha^2 \sigma}{\lambda} \tag{2.12}$$

Assuming the cold junction temperature T_2 to be 300 K, the efficiency is plotted against temperature difference for a range of values of Z in Figure 2.4. ϕ_p does not differ from ϕ_{max} unless Z is large and the generator operates over a large temperature difference. It is apparent that for optimum power generating efficiency the factor $Z\bar{T}$ should be maximized, i.e., a high value of Z should be obtained over as wide a temperature range as possible.

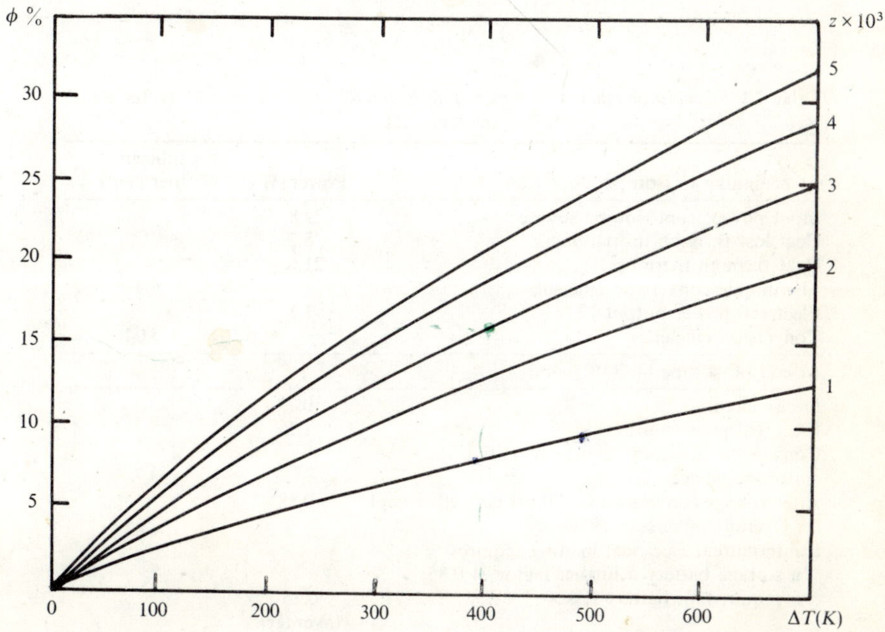

Figure 2.4 Generator efficiency against temperature difference for a range of Z values.

In actual device application current dependent effects should be taken into account and the appropriate Thomson coefficient included in the expression for heat input.[21] Allowance must be made for heat losses in the system. These include thermal conduction losses away from the heat source, thermal conduction losses between source and module, between module and heat sink, flue losses (in the case of a burner heat source), decay of the source (in the case of a radioactive source) and degradation effects in the d.c.–to–d.c. converter. Under matched load conditions, the heat extracted from the source is given by:

$$Q = Q_m + Q_s \tag{2.13}$$

where

$$Q_m = 2nm\lambda \Delta T[1 + \tfrac{1}{2}ZT_1(1 - \tfrac{1}{4}\eta_c)] \tag{2.14}$$

is the heat extracted by the module which has n couples, each with p- and n-type arm of thermal conductivity λ and area–to–length ratio m, and Q_s = heat lost from source by other means. ΔT is the temperature difference across the module, T_1 the hot junction temperature. The system efficiency, ϕ_s, is then given by:[22]

$$\phi_s = \frac{\frac{1}{2}\eta_c}{\left(1 - \frac{1}{4}\eta_c + \frac{2}{ZT_1}\right)}\left(1 - \frac{Q_s}{Q}\right) \qquad 2.15$$

In Table 2.1 an analysis is given of the various losses for a UKAEA 'RIPPLE' generator. Although the module conversion efficiency is almost 7 per cent, the overall end of life system efficiency is only 2.5 per cent.

Table 2.1 *Generator efficiency: typical figures for RIPPLE generators* (After Penn, reference 22)

At beginning of isotope life	Power (W)	Efficiency (per cent)
Input power from isotope source	29.8	
Heat lost through insulation	8.2	
Heat through thermopile	21.6	
Thermopile conversion efficiency		6.9
Electrical power output	1.5	
Conversion efficiency		5.04
At end of isotope life (10 years)		
Decay factor	0.73	
Electrical power output	1.1	
Conversion efficiency relative to initial isotope power		3.7
After voltage conversion (at 80 per cent efficiency)	0.88	
i.e. Overall conversion efficiency		2.95
If intermittent electrical loading requires a storage battery a further factor of 0.85 is required for battery losses	0.75 (average)	
i.e. Overall efficiency		2.5
Total useful electrical energy over 10 years	60 kWh	

2.3 THERMOELECTRIC REFRIGERATION AND THE COEFFICIENT OF PERFORMANCE

If a source of d.c. power is substituted for the load resistance (Figure 2.1, position 3), the thermoelectric device can be used to pump heat from T_1 to T_2, and if $T_1 < T_2$ the device operates as a refrigerator. As in the case of a thermoelectric generator the operation of a refrigerator depends solely upon the properties of the thermocouple material, expressed in terms of the parameter Z, and on the temperatures T_1 and T_2. Thermoelectric refrigerators are designed to operate at either maximum coefficient of performance (COP) or at maximum heat pumping (HP), with coefficient of performance, ε, defined by:

$$\varepsilon = \frac{\text{cooling power}}{\text{input power}} \qquad 2.16$$

The current $I_{max(COP)}$ for maximum coefficient of performance is given by:

$$I_{max(COP)} = \frac{\lambda'(T_2 - T_1)}{\alpha \bar{T}} \sqrt{1 + Z_c \bar{T}} + 1 \qquad 2.17$$

and $$\varepsilon_{max} = \eta' \gamma' \qquad 2.18$$

where $$\eta' = T_1/(T_2 - T_1) \qquad 2.19$$

$$\gamma' = \frac{\sqrt{1 + Z_c \bar{T}} - \dfrac{T_2}{T_1}}{\sqrt{1 + Z_c \bar{T}} + 1} \qquad 2.20$$

and the variation of ε_{max} with $Z_c \bar{T}$ and ΔT is displayed in Figure 2.5. The current for maximum heat pumping $I_{max(HP)}$ is given by:

$$I_{max(HP)} = \frac{\alpha T_1}{R} \qquad 2.21$$

with the maximum temperature differential that can be achieved by a single

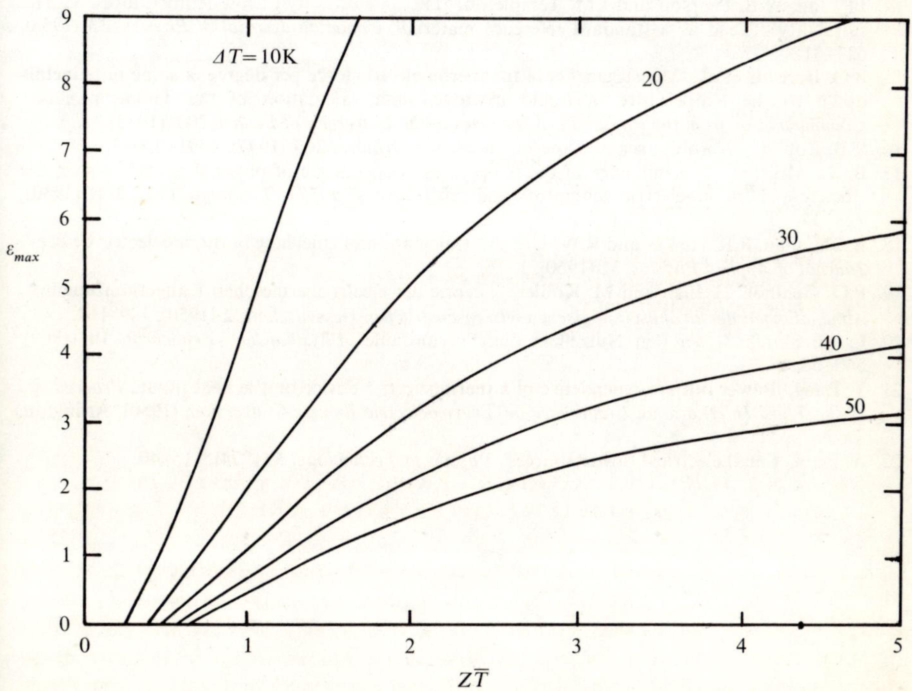

Figure 2.5 Coefficient of performance (optimized for working voltage and geometry) against dimensionless figure of merit; hot junction temperature 300 K.

stage device given by:

$$\Delta T_{max}=(T_2-T_1)_{max}=\tfrac{1}{2}Z_c T_1^2 \qquad 2.22$$

REFERENCES

1. A.F. Ioffe, *Semiconductor thermoelements and thermoelectric cooling*, Infosearch, London, 1957.
2. I.B. Cadoff and E. Miller, (Eds) *Thermoelectric Materials and Devices*, Reinhold, New York, 1959.
3. H.J. Goldsmid, 'Applications of Thermoelectricity', Methuen Monograph, London, 1960.
4. P.E. Gray, *The Dynamic Behavior of Thermoelectric Devices*, Wiley, New York, 1960.
5. R.W. Ure and R.R. Heikes, (Eds) *Thermoelectricity; Science and Engineering*, Interscience, London, 1961.
6. A.I. Burshteyn, *Semiconductor Thermoelectric Devices*, Temple Press, London, 1964.
7. P.J. Bateman, 'Thermoelectric Power Generation', *Contemporary Physics*, **2** (1960), 302–11.
8. F.D. Rosi, E.F. Hocking and N.E. Lindenblad, 'Semiconducting materials for thermoelectric power generation', *RCA Review*, **22** (1961), 82–121.
9. F.D. Rosi, 'Thermoelectricity and Thermoelectric Power Generation', *Solid State Electronics*, **11** (1968), 833–68.
10. D.A. Wright, 'Materials for direct conversion thermoelectric generators' *Metallurgical Reviews*, **15** (1970), 147–60.
11. D.M. Rowe, 'Thermoelectric Power Generation', *Proceedings of Institution of Electrical Engineers*, **125**(11R) (1978), 1113–36.
12. C.M. Bhandari and D.M. Rowe, 'Si–Ge as high temperature thermoelectric materials', *Contemporary Physics*, **21** (3) (1980), 219–42.
13. See for example S.R. de Groot, *Thermodynamics of Irreversible Processes*, North Holland Publishing Co., Amsterdam, 1952.
14. J.P. Jan, W.B. Pearson and I.M. Templeton, 'Thermoelectricity at low temperatures. V. The suitability of lead as a standard reference material', *Canadian Journal of Physics*, **33** (1958), 627–31.
15. B.G. Borelius et al., 'Measurements of the thermoelectric force per degree of some pure metals down to the temperature of liquid hydrogen and calculation of the Thomson effect', *Communication from the Physical Laboratories of the University of Leiden*, 206, (1935).
16. R.B. Roberts, 'Absolute scale of thermoelectricity', *Nature*, **265** (1977), 5591, 226–7.
17. B. Ya Moizhes, 'The influence of the temperature dependence of physical parameters on the efficiency of thermoelectric generators and refrigerators', *Fizika Tverdoga Tela*, **2**(4) (1980), 728–37.
18. B. Sherman, R.R. Heikes and R.W. Ure, Jr, 'Calculation of efficiency of thermoelectric devices', *Journal of Applied Physics*, **31** (1960), 1–16.
19. P.O. Gehlhoff, E. Justi and M. Kohler, 'Theorie der electrochermeschen Kalteergonzangung', *Abhandlungen der Braunschweigischen wissenschaftlichen Gessellschaft*, **2** (1950), 149–164.
20. E. Altenkirch, 'Uber den Nutzeffekt der Thermosaule', *Physikalische Zeitschrift*, **10** (1909), 560–68.
21. V. Raag, 'Power output dependence of a thermoelectric device on the heat input', *Proceedings of the Third International Conference on Thermoelectric Energy Conversion*, (1980), Arlington, Texas, pp. 75–8.
22. A. Penn, 'Small electrical power sources', *Physics in Technology*, **5** (1974), 115–40.

3

Elementary Solid State Theory and Thermoelectric Transport Equations

3.1 INTRODUCTION

THIS chapter deals with the elementary concepts and transport equations which are relevant to the study of thermoelectrics. Solid state theory enables the thermoelectric figure of merit to be represented in terms of a few basic parameters, the behaviour of which can be predicted to some extent. A rigorous calculation of the thermoelectric figure of merit requires a detailed knowledge of the energy band structure and a reasonably good understanding of the various mechanisms which scatter charge carriers, together with some knowledge of phonon dispersion relationships and scattering mechanisms. A quantitative analysis which includes all the relevant factors could be attempted but the equations soon become unwieldy. In many instances it is possible to simplify the mathematics by making drastic assumptions in the model without it becoming too unrealistic. The model can then be refined as the need arises.

3.2 BASIC CONCEPTS AND THE FERMI LEVEL

In a metal at 0 K all states up to a certain level (referred to as the Fermi level) are filled with electrons, while states above this are empty. At a non-zero temperature, energy states in a range of about $4kT$ around the Fermi level are partially filled. The probability of finding an electron at the Fermi level is one half, while for other states the probability can be related to the energy difference between this state and the Fermi level. In a semiconductor, the concept of a Fermi level is not as simple. However, the probability of a state being occupied depends upon the energy difference of the state and the Fermi level in exactly the same way as in metals. A brief description of the variation of the Fermi level with temperature and carrier concentration is given in the following pages.

Determination of the Fermi level.[1] A general model of a semiconductor comprises a conduction band and a valence band separated by an energy gap (Figure 3.1). In impurity semiconductors localized states exist in the gap region and correspond to donor or accepter impurities. Taking the zero of energy at the bottom of the conduction band one defines the energies corresponding to donor level and accepter level as $-\Delta E_b$ and $-\Delta E_a$. The valence band edge (top of the valence band) lies at an energy $-E_g$.

The density of states $D(E)$ near the conduction band edge (for $E>0$, where E is the energy of the charge carriers) and valence band edge (for $E<-E_g$) is assumed to be parabolic with:

$$D(E)=\frac{4\pi}{h^3}(2m_e^*)^{\frac{3}{2}}E^{\frac{1}{2}} \qquad E>0 \qquad 3.1$$

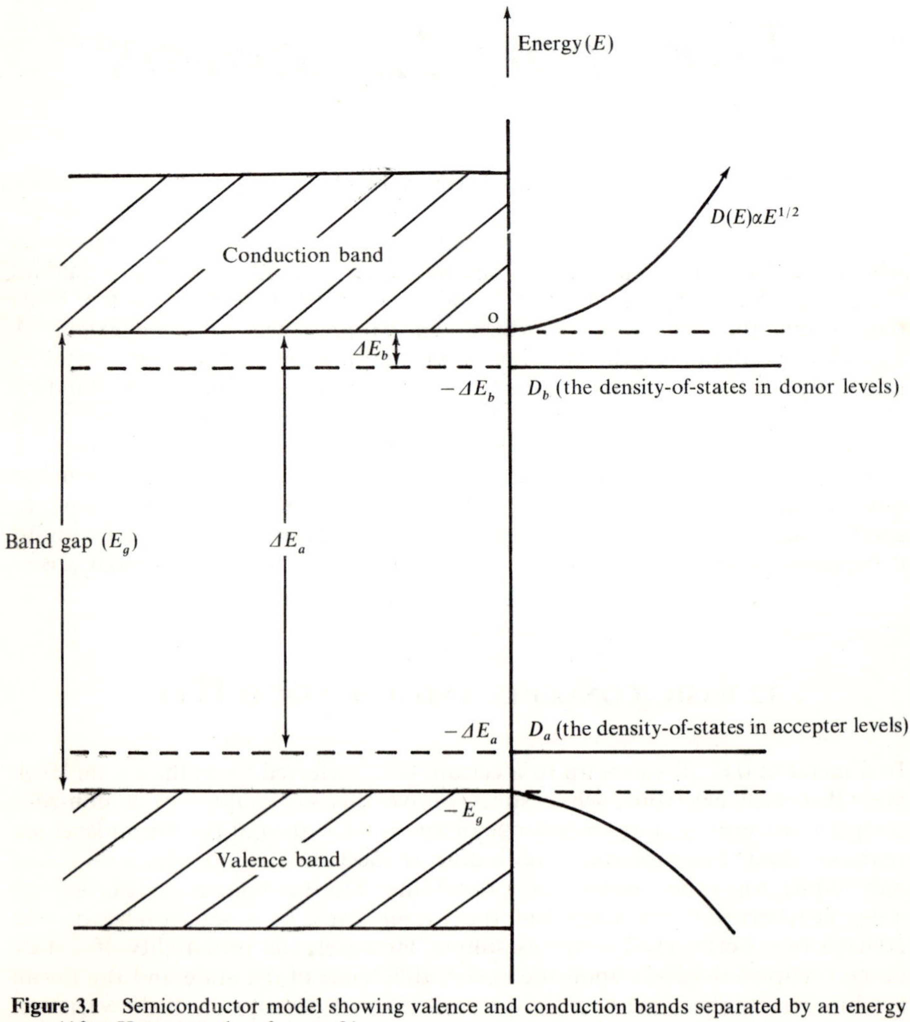

Figure 3.1 Semiconductor model showing valence and conduction bands separated by an energy gap. (After Hutner et al., reference 2.)

$$D(E)=\frac{4\pi}{h^3}(2m_h^*)^{\frac{3}{2}}(|E|-E_g)^{\frac{1}{2}} \qquad E<-E_g \qquad 3.2$$

m_e^*, m_h^* refer to electron and hole density of states effective masses, respectively. In the gap region, $-E_g<E<0$, $D(E)$ at each impurity level can be written in terms of Dirac δ functions.

$$D(E)=D_a\delta(E+\Delta E_a)+D_b\delta(E+\Delta E_b) \qquad 3.3$$

D_a, D_b are number of states per m^3 in accepter and donor levels.

For many purposes a simple model with one band and one impurity level is used.[2] For a conduction band and a donor level the Fermi level can be determined (see Appendix, Section 3.6) by solving the following equation:

$$\frac{2P}{\sqrt{\pi}}F_{\frac{1}{2}}(\xi)=\frac{D_b e^{-\Delta E_b/kT}}{e^{-\Delta E_b/kT}+e^{\xi}} \qquad 3.4$$

here $$P=\frac{2(2\pi m_e^* kT)^{\frac{3}{2}}}{h^3}$$

and $$F_{\frac{1}{2}}(\xi)=\int_0^\infty \frac{x^{\frac{1}{2}}dx}{e^{(x-\xi)}+1} \qquad 3.5$$

where $\xi=E_F/kT$ is the reduced (dimensionless) Fermi level. A schematic plot of ξ versus T for different donor atom concentrations is shown in Figure 3.2.

However, a single-band model creates difficulties as the Fermi level diverges without limit from the band edge. A two-band model removes this difficulty, and the equation one must now solve (for example, with two bands and a donor level) is given as:

$$\frac{2P}{\sqrt{\pi}}F_{\frac{1}{2}}(\xi)=\frac{D_b e^{-\Delta E_b/kT}}{e^{-\Delta E_b/kT}+e^{\xi}}+\frac{2Q}{\sqrt{\pi}}F_{\frac{1}{2}}(-\xi-\xi_g) \qquad 3.6$$

where $$Q=\frac{2(2\pi m_h^* kT)^{\frac{3}{2}}}{h^3}$$

and $\xi_g=E_g/kT$ the dimensionless reduced energy band gap.

Results of calculation for germanium are shown in Figure 3.3 where the Fermi level is plotted against temperature for various donor atom concentrations. It is convenient to quote energies in electron volts, with $1\,\text{eV}=0.1603\times10^{-18}$ J. One can observe that instead of diverging indefinitely from the band edge the Fermi level now approaches the intrinsic value. This calculation does not consider the variation of the energy gap and effective mass with temperature. The temperature variation of E_g in germanium is fairly well known and this can be taken into consideration in calculating the Fermi level. However, the variation of the effective mass and impurity level position are not so well understood and in many simple calculations the effective mass is assumed to be independent of temperature.

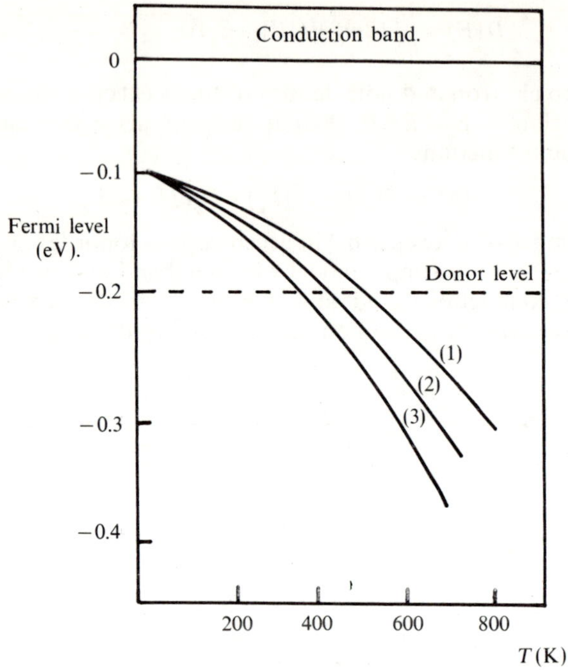

Figure 3.2 Temperature dependence of Fermi level in simple *n*-type semiconductors. (After Hutner et al., reference 2.)

3.3 HEAT CONDUCTION

Thermal conduction in semiconductors[3,4] is mainly due to the lattice contribution (λ_L), while in heavily doped materials the electronic or hole contribution (λ_p) may become significant. In the range of intrinsic conduction the bipolar contribution (λ_b) to thermal conductivity should also be taken into consideration. In general:

$$\lambda = \lambda_{lattice} + \lambda_{polar} + \lambda_{bipolar}$$
$$= \lambda_L + \lambda_p + \lambda_b \qquad 3.7$$

The thermal vibrations of the lattice are quantized and the resulting quanta (phonons) flow down the temperature gradient and constitute a thermal current. These phonons, as they flow, may be scattered by other phonons, by disorder in the lattice, and by boundaries. Carriers may also cause an appreciable scattering of phonons in heavily doped semiconductors. The presence of free carriers affects the total thermal conductivity in two ways: electrons act as carriers of heat and also as scatterers of phonons.

Solid solution alloys of various semiconductors have been investigated in recent years because of the reduction in thermal conductivity which accompanies alloying. The dependence of the lattice thermal conductivity on alloy

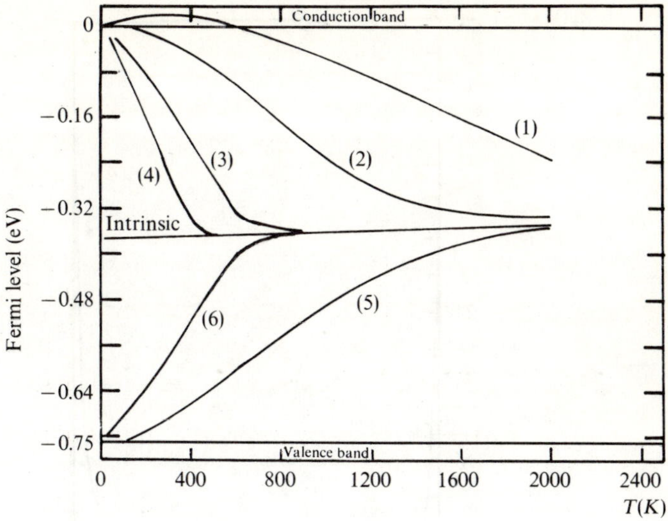

Carrier concentration values (m^{-3})
(1) $D_b = 10^{26}$ (2) $D_b = 10^{25}$ (3) $D_b = 10^{23}$
(4) $D_b = 10^{21}$ (5) $D_a = 10^{25}$ (6) $D_a = 10^{23}$
$\Delta E_b = 0.005$eV below the conduction band.
$\Delta E_a = 0.01$eV above the valence band.
$E_g = 0.75$eV, $m_h^*/m_e^* = 1.2$.

Figure 3.3 Fermi level as a function of temperature and doping level in germanium, a two-band model with a donor or accepter level. (After Hutner et al., reference 2.)

disorder is shown[5] in Figure 3.4 for the GaAs–InAs system. In many systems, alloy disorder scattering reduces the thermal conductivity by almost an order of magnitude. The role of boundary scattering in limiting the phonon mean free path has usually been considered as a low temperature phenomenon where, in the absence of other phonon scattering mechanisms, boundary scattering effects are observed as the phonon mean-free path approaches the sample dimensions. However, in highly disordered systems, because a large proportion of heat is carried by low frequency phonons,[6] boundary scattering effects can manifest themselves in the region of the Debye temperature and above, and in principle lead to a reduction in the lattice thermal conductivity.

3.3.1 Lattice Thermal Conductivity—Relaxation Time Approach

From simple considerations based on the kinetic theory of gases one can obtain the following expression for the lattice thermal conductivity in terms of the phonon mean-free-path (l), phonon velocity (v_s) and phonon heat capacity per unit volume (C):

$$K_L = 1/3\, C v_s l \qquad 3.8$$

The phonon mean free path is determined by the extent to which phonons can be scattered by other phonons, lattice disorder, free cha e carriers, and by

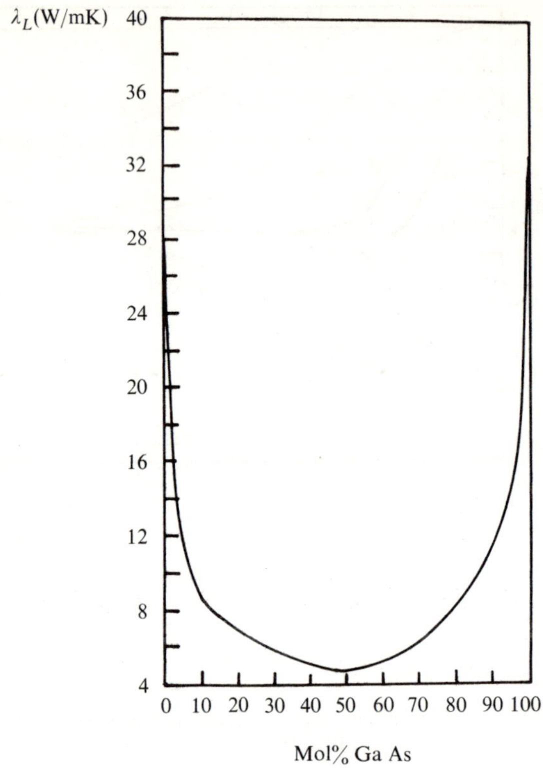

Figure 3.4 Dependence of lattice thermal conductivity on alloy disorder in GaAs–InAs alloy system. (After Abrahams et al., reference 5.)

crystal-or-grain boundaries. It has been useful and customary to work out formulae in terms of a certain characteristic for phonons—the relaxation time, τ. The relaxation time may depend upon the phonon frequency and temperature as well as the crystal structure of the material.

The scattering of phonons by other phonons is due to anharmonic lattice interactions which bring about a coupling between different phonons. The three-phonon process, defined by $\vec{q}_1+\vec{q}_2=q_3+G$, is important. Here the $\vec{q}$s refer to phonon wave vectors and $\vec{G}$ is the reciprocal lattice vector. For the case $\vec{G}=0$ the total crystal momentum is conserved and the process is called a normal or N-process. Other values of $G(\neq 0)$ correspond to Umklapp or U-processes where the momentum is not conserved.

The frequency and temperature dependence of the relaxation time depend strongly on the actual phonon branch and the dispersion in the phonon spectrum, and in order to obtain simple workable formulae a number of simplifying assumptions have to be introduced. The usual approach consists in assuming a relaxation time approximation in the Boltzmann equation and using an isotropic Debye-like phonon spectrum consisting of one average acoustic branch. Herring[8] has considered the crystal class explicitly in obtaining an expression for the relaxation time which, for the three-phonon normal process, is given by:

$$\tau^{-1}(q) \propto q^{y} \qquad 3.9$$

the exponent y being determined by crystal symmetry; for example, $y=2$ for longitudinal-acoustic modes in a simple-cubic crystal. Callaway's phenomenological model[9] achieved great success in obtaining the lattice thermal conductivity of a large number of substances. Phonon scattering by lattice-disorder, charge-carriers and boundaries are other mechanisms which must be considered in a detailed analysis.

Usually one considers the disorder scattering arising solely due to mass difference:

$$\tau_{dis}^{-1} = \frac{\Gamma v_s \Omega_0}{4\pi} \left(\frac{kT}{\hbar v_s}\right)^4 z^4 \qquad 3.10$$

Here $z = \frac{\hbar\omega}{kT}$ where ω is the phonon frequency, Ω_0 is the volume per atom. Γ is the disorder parameter and is given by $\Gamma = \Sigma_i f_i \left(1 - \frac{M_i}{M}\right)^2$ where $M = \Sigma_i f_i M_i$, f_i being the fraction of atoms of mass M_i.

The relaxation time due to scattering by boundaries to a reasonably good approximation is given by L/v_s where L is a length similar to the linear crystal-or-grain-dimensions and is referred to as the Casimir length.

The scattering of phonons by electrons (holes) has been discussed by Ziman.[11] For low frequency phonons and large concentrations of carriers one can write:[12]

$$\tau_{ep}^{-1} = B'z \qquad 3.11$$

with

$$B' = \frac{\varepsilon_1^2 m^{*2} kT}{2\pi \hbar^4 \rho v_s} \qquad 3.12$$

where ε_1 is the deformation potential, m^* the density of states effective-mass and ρ is the density. It is therefore reasonable to proceed with the assumption that τ_{ep}^{-1} is proportional to z.

3.3.2 Sintered Materials

A simple analysis[12,13] of the thermal conductivity of sintered semiconductor alloys with heavy doping is of particular interest in thermoelectrics. The Klemens–Callaway theory gives for the lattice component of thermal conductivity:

$$\lambda_L = \frac{k}{2\pi^2 v_s} \left(\frac{kT}{\hbar}\right)^3 \left(I_1 + \frac{I_2^2}{I_3}\right) \qquad 3.13$$

The first term in brackets refers to the contribution from momentum-destroying processes and treating normal processes at par with them. The second bracketed term refers to a correction due to the normal processes.

At high temperatures one can replace the Planck distribution $(e^z - 1)^{-1}$ by

z^{-1} and one obtains:

$$I_1=\int_0^{\Theta/T}\tau_c z^2\,dz \qquad 3.14$$

$$I_2=\int_0^{\Theta/T}\frac{\tau_c}{\tau_N}z^2\,dz \qquad 3.15$$

$$I_3=\int_0^{\Theta/T}\frac{1}{\tau_N}\left(1-\frac{\tau_c}{\tau_N}\right)z^2\,dz \qquad 3.16$$

where Θ is a phonon characteristic temperature. τ_c is the combined relaxation time given by:

$$\tau_c^{-1}=\sum_i \tau_i^{-1} \qquad 3.17$$

and where τ_i refers to the relaxation time for ith scattering mechanism. Further, if one assumes the simple relationship between τ_N and τ_U:[13]

$$\tau_N^{-1}=k_0\tau_u^{-1}=k_0 A_u z^2 \qquad 3.18$$

then one can express A_u in terms of the thermal conductivity λ_0 of the corresponding, and notional, perfect large single crystal (three-phonon scattering being the only scattering mechanism), and λ_0 is given by:

$$\lambda_0=\frac{k}{2\pi^2 v_s}\left(\frac{kT}{\hbar}\right)^3\frac{\Theta}{A_u T}\,\frac{1+5k_0/9}{1+k_0} \qquad 3.19$$

For an undoped sintered semiconductor alloy it can readily be shown that:

$$\frac{\lambda_L}{\lambda_0}=(1+5k_0/9)^{-1}\left\{L_2+\frac{k_0L_4^2}{(1+k_0)AL_8+L_6+(1+k_0)v_sL_4}\right\} \qquad 3.20$$

where

$$L_n(A,C)=\int_0^1\frac{x^n\,dx}{Ax^4+x^2+C} \qquad 3.21$$

$$C=\frac{2\pi^2\lambda_0 v_s^2}{k\omega_D^3 L(1+5k_0/9)} \qquad 3.22$$

$$A=\frac{\pi\lambda_0\Gamma\Omega_0\omega_D}{2v_s^2k(1+5k_0/9)} \qquad 3.23$$

The second term in the expression of thermal conductivity is usually small and can be neglected in calculations.

For doped material one has to include the effect of phonon–electron scattering. Assuming a simple relaxation time as discussed in the previous section, the thermal conductivity for a doped-sintered semiconductor alloy is given by:

$$\frac{\lambda_L}{\lambda_0}=(1+5k_0/9)^{-1}L_2(A,B,C) \qquad 3.24$$

where $$L_2(A,B,C)=\int_0^1 \frac{x^2\,dx}{Ax^4+x^2+Bx+C} \tag{3.25}$$

and $$B=\frac{2\pi^2\lambda_0\hbar B'}{\omega_D k^2 T(1+5k_0/9)\frac{\omega_D}{v_s}} \tag{3.26}$$

Based on this simple analysis the thermal conductivity of Si–Ge alloys has been investigated in detail.[12-14] The results of a typical calculation of the variation of thermal conductivity with doping and grain size is shown in Figure 3.5.

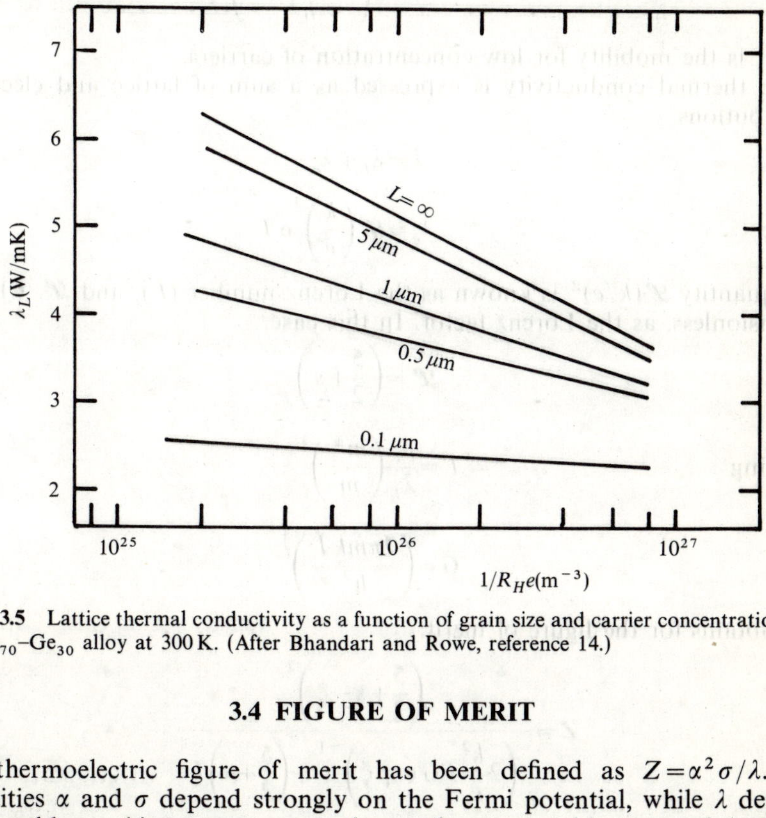

Figure 3.5 Lattice thermal conductivity as a function of grain size and carrier concentration in p-type Si_{70}–Ge_{30} alloy at 300 K. (After Bhandari and Rowe, reference 14.)

3.4 FIGURE OF MERIT

The thermoelectric figure of merit has been defined as $Z=\alpha^2\sigma/\lambda$. The quantities α and σ depend strongly on the Fermi potential, while λ depends only weakly on this parameter. α and σ can be expressed in terms of the Fermi potential using classical statistics (low carrier concentration) or Fermi–Dirac statistics (high carrier concentration).

3.4.1 Classical Statistics

First we shall discuss the situation assuming classical statistics for electrons (or holes). In a simple model it is usual to assume that the density of states is proportional to $E^{\frac{1}{2}}$ and that the relaxation time τ_e for charge carriers can be expressed in the form $\tau_e \alpha E^s$; s is a constant and is referred to as the scattering parameter. It can be shown that:[15]

$$\alpha = \mp \frac{k}{e}\left(\frac{5}{2} + s - \xi\right) \qquad 3.27$$

where − and + signs refer to electrons and holes respectively.

The electrical conductivity[16] is related to mobility μ and concentration of charge carriers n by:

$$\sigma = ne\mu_c \qquad 3.28$$

where

$$n = 2\left(\frac{2\pi m^* kT}{h^2}\right)^{\frac{3}{2}} e^{\xi} \qquad 3.29$$

and μ_c is the mobility for low concentration of carriers.

The thermal conductivity is expressed as a sum of lattice and electronic contributions.

$$\lambda = \lambda_L + \lambda_e \qquad 3.30$$

where

$$\lambda_e = \mathscr{L}\left(\frac{k}{e}\right)^2 \sigma T \qquad 3.31$$

The quantity $\mathscr{L}(k/e)^2$ is known as the Lorenz number (L), and $\mathscr{L}$, which is dimensionless, as the Lorenz factor. In this case:

$$\mathscr{L} = \left(\frac{5}{2} + s\right) \qquad 3.32$$

Defining

$$F = \frac{\mu_c}{\lambda_L}\left(\frac{m^*}{m}\right)^{\frac{3}{2}} \qquad 3.33$$

and

$$G = \left(\frac{2\pi mkT}{h^2}\right)^{\frac{3}{2}} \qquad 3.34$$

one obtains for the figure of merit:

$$Z = \frac{\left(\frac{5}{2} + s - \xi\right)^2}{\left(2\frac{k^2}{e} FG \exp\xi\right)^{-1} + \left(\frac{5}{2} + s\right)T} \qquad 3.35$$

The figure of merit changes with the Fermi level. The Fermi level for which Z is a maximum can be obtained by setting $dZ/d\xi$ equal to zero. This gives:

$$\xi + 4\left(\frac{5}{2} + s\right)\frac{k^2}{e} FGT \exp\xi = \frac{1}{2} + s \qquad 3.36$$

In general, the lattice component of thermal conductivity is much larger than the electronic component. One can therefore neglect the second term to obtain a simple relation for the optimum value of Fermi level:

$$\xi_{opt} = \tfrac{1}{2} + s \qquad 3.37$$

For acoustic lattice scattering and ionized-impurity scattering s is equal to $-\frac{1}{2}$ and $\frac{3}{2}$ respectively. This gives a value of optimum Fermi level which lies between 0 and 2. Inclusion of the electronic term (neglected above) will have the effect of reducing ξ_{opt}.

Further, if the expression $Z=\alpha^2\sigma/\lambda$ is differentiated with respect to α at constant temperature and the result equated to zero, one obtains a condition for a maximum of Z as:

$$\alpha\left(\frac{d}{d\alpha}\log\sigma\right)=-2\left[1+\frac{\lambda_e}{\lambda_L}\left(1-\frac{1}{2}\frac{d(\log L)}{d(\log\alpha)}\right)\right] \qquad 3.38$$

Using classical statistics:

$$\frac{d(\log\sigma)}{d\alpha}=\frac{e}{k} \qquad 3.39$$

for all scattering mechanisms.

The optimum value of the Seebeck coefficient is given by:

$$\alpha=-2\left(\frac{k}{e}\right)\left[1+\frac{\lambda_e}{\lambda_L}\right]=-172\left[1+\frac{\lambda_e}{\lambda_L}\right]\mu V/K \qquad 3.40$$

Substituting this value of α in the expression of the figure of merit one obtains:

$$Z_m T=\frac{4\left(\frac{k}{e}\right)^2}{L}\frac{\lambda_e}{\lambda_L}\left(1+\frac{\lambda_e}{\lambda_L}\right) \qquad 3.41$$

In this model higher values of $Z_m T$ will be obtained when the electronic contribution to thermal conductivity is a significant fraction of the total thermal conductivity. For λ_e/λ_L around 0.5, $\alpha_{opt}=250\,\mu V/K$.

These simple considerations are only illustrative. A detailed numerical investigation is essential to obtain more realistic estimates and this will form the theme of the next section. As indicated earlier, the optimum Fermi level lies around zero and this corresponds to a high concentration of charge carriers. Under these circumstances use of the simple models described may be inappropriate for a number of reasons:

1. At high values of carrier concentration the usual $\tau\propto E^s$ type relationship is inadequate as the relaxation time may depend on the carrier concentration as well as on the carrier energy.
2. Use of classical statistics is not justified except at very low carrier concentrations.
3. Impurity-level broadening occurs which results in a possible overlap with the conduction or valence bands.
4. Ionized-impurity scattering of charge carriers will be significant at high concentration, and the usual Conwell–Weisskopf[17] formula normally used for the description of mobility may be an oversimplification.

However, these simple models are important in their own right and are particularly valuable for a qualitative understanding of the various aspects of the problem.

3.4.2 Fermi–Dirac Statistics

When Fermi–Dirac statistics are used the Seebeck coefficient is given by:

$$\alpha = \mp \frac{k}{e}(\delta - \xi) \quad 3.42$$

where
$$\delta = \frac{(s+\frac{5}{2})F_{s+\frac{3}{2}}(\xi)}{(s+\frac{3}{2})F_{s+\frac{1}{2}}(\xi)} \quad 3.43$$

and the Fermi integral is defined as:

$$F_l(\xi) = \int_0^\infty \frac{x^l \, dx}{1+e^{(x-\xi)}} \quad 3.44$$

The Lorenz number $L = \left(\frac{k}{e}\right)^2 \mathscr{L}$ 3.45

where
$$\mathscr{L} = \frac{(s+\frac{7}{2})F_{s+\frac{5}{2}}(\xi)}{(s+\frac{3}{2})F_{s+\frac{1}{2}}(\xi)} - \left\{\frac{(s+\frac{5}{2})F_{s+\frac{3}{2}}(\xi)}{(s+\frac{3}{2})F_{s+\frac{1}{2}}(\xi)}\right\}^2 \quad 3.46$$

For $-\xi \gg 1$, i.e. the classical limit, both δ and $\mathscr{L}$ are reduced to $(s+\frac{5}{2})$.

Variation of α with reduced Fermi level ξ is shown in Figure 3.6[18] for three different types of carrier scattering: acoustic lattice scattering ($s = -\frac{1}{2}$), optical phonon scattering ($s = \frac{1}{2}$) and ionized impurity scattering ($s = \frac{3}{2}$). These carrier scattering processes are discussed in greater detail in Chapter 4.

In a number of situations it is useful to express the electrical conductivity in a form that is convenient in the calculation of the figure of merit for various scattering mechanisms. It is customary to define a relaxation time for charge carriers which varies as the Sth power of the carrier energy. Then:

$$\tau(E) = aE^S \quad 3.47$$

where a is independent of energy although it may depend upon temperature. The charge carrier mobility can now be written as:

$$\mu = \frac{e}{m_c}\langle\tau\rangle \quad 3.48$$

Here m_c is the conductivity (inertial) effective-mass and $\langle\tau\rangle$ is the average relaxation time defined as:

$$\langle\tau\rangle = \frac{\int_0^\infty \tau(E)E^{\frac{3}{2}}\left(\frac{\partial f_0}{\partial E}\right)dE}{\int_0^\infty E^{\frac{3}{2}}\left(\frac{\partial f_0}{\partial E}\right)dE} \quad 3.49$$

with
$$f_0(E) = \left[1+\exp\left(\frac{E}{kT} - \xi\right)\right]^{-1} \quad 3.50$$

It follows that
$$\mu = \frac{2}{3}a(kT)^s\left(\frac{e}{m_c}\right)\frac{F_{s+\frac{1}{2}}(\xi)}{F_{\frac{1}{2}}(\xi)} \quad 3.51$$

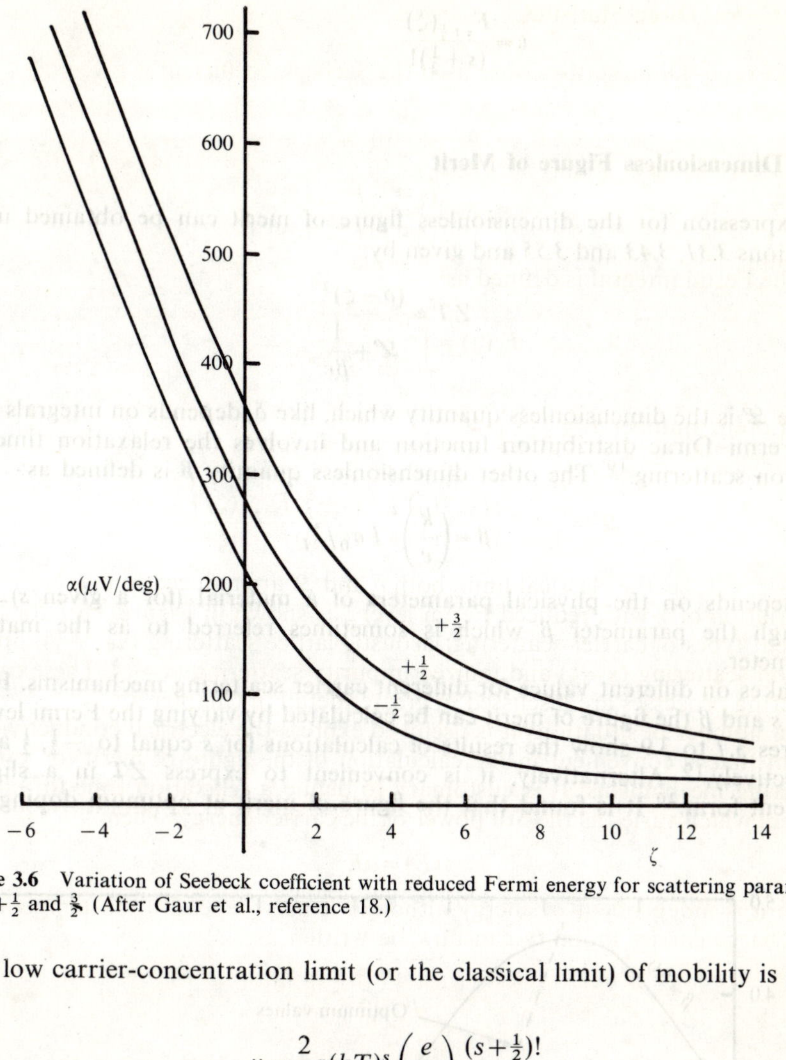

Figure 3.6 Variation of Seebeck coefficient with reduced Fermi energy for scattering parameters $-\frac{1}{2}$, $+\frac{1}{2}$ and $\frac{3}{2}$. (After Gaur et al., reference 18.)

The low carrier-concentration limit (or the classical limit) of mobility is given by:

$$\mu_c = \frac{2}{3} a (kT)^s \left(\frac{e}{m_c}\right) \frac{(s+\frac{1}{2})!}{\frac{1}{2}!} \qquad 3.52$$

Hence

$$\mu = \frac{\pi^{\frac{1}{2}}}{2} \frac{F_{s+\frac{1}{2}}(\xi)}{(s+\frac{1}{2})!\,F_{\frac{1}{2}}(\xi)} \mu_c \qquad 3.53$$

The carrier concentration in a single-band is given by:

$$n = \frac{4}{\sqrt{\pi}} (2\pi m^* kT/h^2)^{\frac{3}{2}} F_{\frac{1}{2}}(\xi) \qquad 3.54$$

Using these equations one can write:

$$\sigma = ne\mu = \sigma_0 \varepsilon \qquad 3.55$$

where

$$\sigma_0 = 2(2\pi m^* kT/h^2)^{\frac{3}{2}} e \mu_c \qquad 3.56$$

and
$$\varepsilon = \frac{F_{s+\frac{1}{2}}(\xi)}{(s+\frac{1}{2})!} \qquad 3.57$$

3.4.3 Dimensionless Figure of Merit

An expression for the dimensionless figure of merit can be obtained using equations *3.31*, *3.43* and *3.55* and given by:

$$ZT = \frac{(\delta - \xi)^2}{\mathscr{L} + \frac{1}{\beta\varepsilon}} \qquad 3.58$$

where $\mathscr{L}$ is the dimensionless quantity which, like δ, depends on integrals over the Fermi–Dirac distribution function and involves the relaxation time for electron scattering.[19] The other dimensionless quantity β is defined as:

$$\beta = \left(\frac{k}{e}\right)^2 T\sigma_0/\lambda_L \qquad 3.59$$

ZT depends on the physical parameters of a material (for a given s) only through the parameter β which is sometimes referred to as the material parameter.

s takes on different values for different carrier scattering mechanisms. For a fixed s and β the figure of merit can be calculated by varying the Fermi level ξ. Figures 3.7 to 3.9 show the results of calculations for s equal to $-\frac{1}{2}$, $\frac{1}{2}$ and $\frac{3}{2}$ respectively.[19] Alternatively, it is convenient to express ZT in a slightly different form.[20] It is found that the figure of merit at optimum doping is a

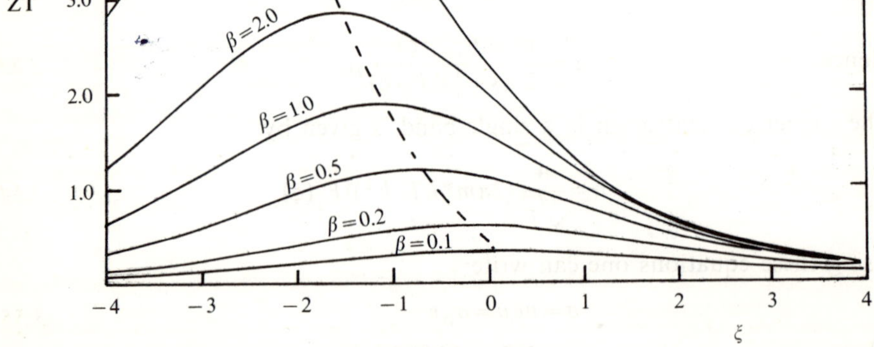

Figure 3.7 Dependence of the dimensionless figure of merit on reduced Fermi energy for $s = -\frac{1}{2}$. (After Chasmar and Stratton, reference 19.)

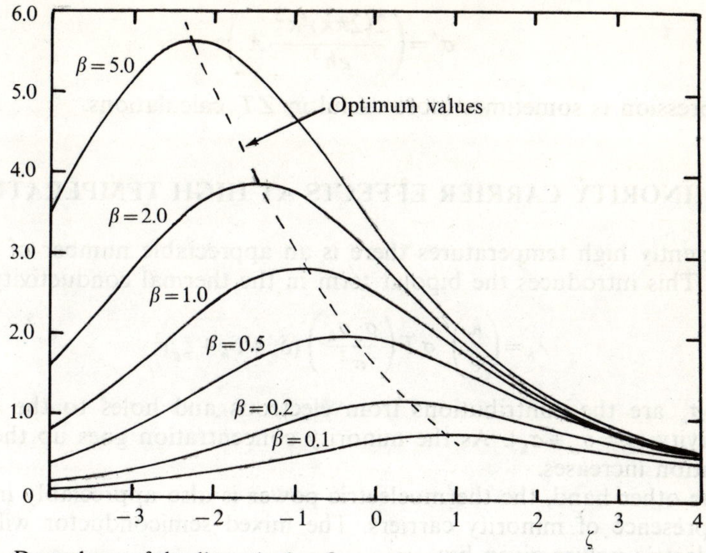

Figure 3.8 Dependence of the dimensionless figure of merit on reduced Fermi energy for $s=+\frac{1}{2}$. (After Chasmar and Stratton, reference 19.)

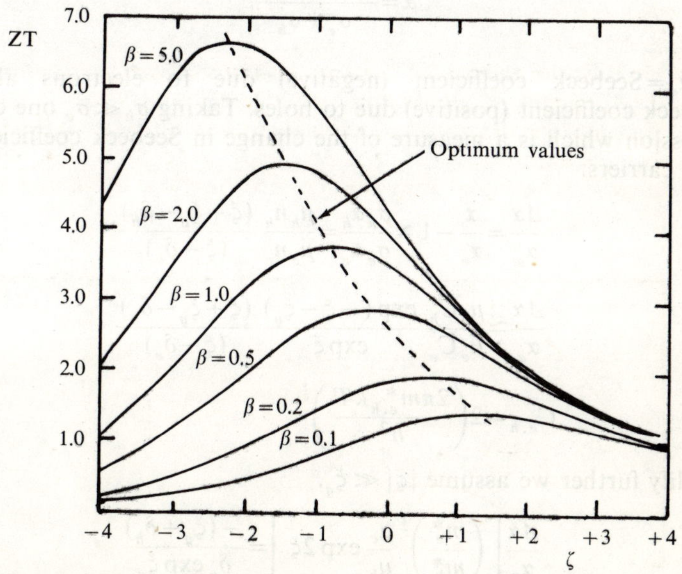

Figure 3.9 Dependence of the dimensionless figure of merit on reduced Fermi energy for $s=+\frac{3}{2}$. (After Chasmar and Stratton, reference 19.)

function of a parameter A' defined as:

$$A' \equiv T^{\frac{5}{2}} m^{*\frac{3}{2}} \mu_c / \lambda_L \qquad 3.60$$

ZT is found to increase monotonically with this parameter. One then writes for the dimensionless electrical conductivity:[19]

$$\sigma' = \left(\frac{k}{e}\right)^2 \frac{T}{\lambda_L} \sigma = \beta \varepsilon \qquad 3.61$$

or
$$\sigma' = \left(\frac{2(2\pi k)^{\frac{3}{2}}k^2}{eh^3}A'\right)\varepsilon \tag{3.62}$$

This expression is sometimes more useful in ZT calculations.

3.5 MINORITY CARRIER EFFECTS AT HIGH TEMPERATURES

At sufficiently high temperatures there is an appreciable number of minority carriers. This introduces the bipolar term in the thermal conductivity.

$$\lambda_b = \left(\frac{k}{e}\right)^2 \sigma T \left(\frac{\sigma_e \sigma_h}{\sigma^2}\right)(\delta_e + \delta_h + \xi_g)^2 \tag{3.63}$$

σ_e and σ_h are the contributions from electrons and holes to the electrical conductivity ($\sigma = \sigma_e + \sigma_h$). As the minority concentration goes up the bipolar contribution increases.

On the other hand, the thermoelectric power is also appreciably influenced by the presence of minority carriers. The mixed semiconductor will have a thermoelectric power given by:

$$\alpha = \frac{\alpha_e \sigma_e + \alpha_h \sigma_h}{\sigma_e + \sigma_h} \tag{3.64}$$

where α_e = Seebeck coefficient (negative) due to electrons alone, and α_h = Seebeck coefficient (positive) due to holes. Taking $\sigma_h \ll \sigma_e$ one can obtain an expression which is a measure of the change in Seebeck coefficient due to minority carriers:

$$\frac{\Delta\alpha}{\alpha_e} = \frac{\alpha}{\alpha_e} - 1 \simeq \frac{\sigma_h \alpha_h}{\sigma_e \alpha_e} = \frac{\mu_h n_h}{\mu_e n_e} \frac{(\xi + \xi_g - \delta_h)}{(\xi - \delta_e)} \tag{3.65}$$

$$\frac{\Delta\alpha}{\alpha_e} \simeq \frac{\mu_h C_h}{\mu_e C_e} \frac{\exp(-\xi - \xi_g)}{\exp \xi} \frac{(\xi + \xi_g - \delta_h)}{(\xi - \delta_e)} \tag{3.66}$$

Here
$$C_{e,h} = 2\left(\frac{2\pi m^*_{e,h} kT}{h^2}\right)^{\frac{3}{2}} \tag{3.67}$$

To simplify further we assume $|\xi| \ll \xi_g$.

Then
$$\frac{\Delta\alpha}{\alpha_e}\left[\left(\frac{m_e^*}{m_h^*}\right)^{\frac{3}{2}} \frac{\mu_e}{\mu_h} \exp 2\xi\right] = \frac{-(\xi_g + \delta_h)}{\delta_e \exp \xi_g} \tag{3.68}$$

where m_e^* and m_h^* are the density of states effective masses. The left hand side of equation 3.68 is plotted as a function of reduced energy gap ξ_g in Figure 3.10 for the case $\delta_e = \delta_h = 2$.

3.6 APPENDIX

At absolute zero the number of electrons per m³ (i.e., the concentration) in the valence-band is equal to N. Let N_d be the concentration of electrons in the

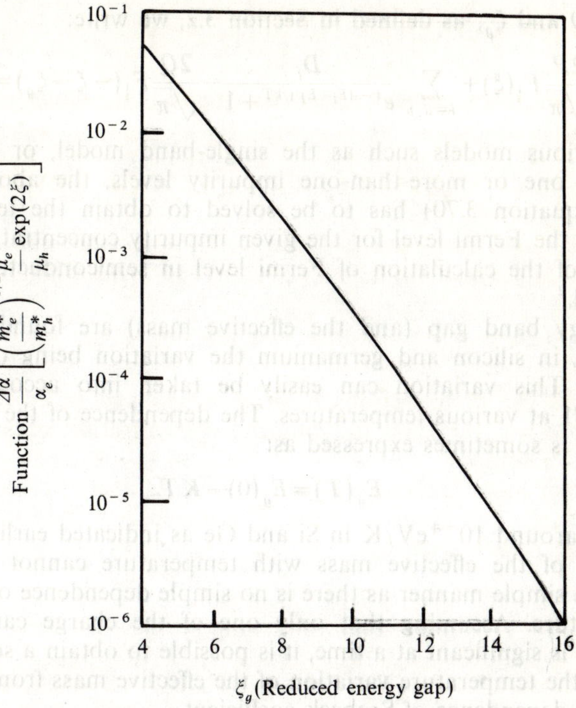

Figure 3.10 The function $\frac{\Delta\alpha}{\alpha_e}\left[\left(\frac{m_e^*}{m_h^*}\right)^{\frac{3}{2}}\frac{\mu_e}{\mu_h}\exp(2\xi)\right]$ versus ξ_g for $\delta_e = \delta_n = 2$.

impurity levels. One can then write:

$$\int_{-\infty}^{\infty} D(E) f_o(E)\,\mathrm{d}E = N + N_d \qquad 3.69$$

where
$$f_o(E) = \left(\exp\left(\frac{E-E_F}{kT}\right)+1\right)^{-1} \qquad 3.70$$

E_F refers to the energy of the Fermi level. The evaluation of the integral can be made easier by dividing the energy scale into three regions: $E>0$, $-E_g<E<0$, and $E<-E_g$.

With the density of states for the three regions as defined in Section 3.2 of this chapter, the above equation (3.69) can be written as:

$$\frac{4\pi}{h^3}(2m_e^*)^{\frac{3}{2}}\int_0^{\infty}\frac{E^{\frac{1}{2}}\mathrm{d}E}{e^{(E-E_F)/kT}+1} + \int_{-E_g}^{0}\frac{[D_a\delta(E+\Delta E_a)+D_b\delta(E+\Delta E_b)]}{e^{(E-E_F)/kT}+1}\mathrm{d}E$$

$$+N - \frac{4\pi}{h^3}(2m_h^*)^{\frac{3}{2}}\int_{E_g}^{\infty}\frac{(E-E_g)^{\frac{1}{2}}\mathrm{d}E}{e^{(E+E_F)/kT}+1} = N + N_d \qquad 3.71$$

Use has been made of the fact that for $E<-E_g$, concentration of electrons at any temperature $= N -$ concentration of holes.

With P, Q and ξ_g, as defined in Section 3.2, we write:

$$\frac{2P}{\sqrt{\pi}}F_{\frac{1}{2}}(\xi)+\sum_{i=a,b}\frac{D_i}{e^{(-\Delta E_i-E_F)/kT}+1}-\frac{2Q}{\sqrt{\pi}}F_{\frac{1}{2}}(-\xi-\xi_g)=N_d \tag{3.72}$$

For the various models such as the single-band model, or the two-bands model with one or more-than-one impurity levels, the above generalized equation (equation 3.70) has to be solved to obtain the temperature dependence of the Fermi level for the given impurity concentration. A detailed description of the calculation of Fermi level in semiconductors is given by Hutner et al.[2]

The energy band gap (and the effective mass) are found to vary with temperature, in silicon and germanium the variation being of the order of 10^{-4}eV/K. This variation can easily be taken into account by solving equation 3.71 at various temperatures. The dependence of the energy gap on temperature is sometimes expressed as:

$$E_g(T)=E_g(0)-KT \tag{3.73}$$

where K is around 10^{-4}eV/K in Si and Ge as indicated earlier.

Variation of the effective mass with temperature cannot be taken into account in a simple manner as there is no simple dependence of effective mass on temperature. Assuming that only one of the charge carrier scattering mechanisms is significant at a time, it is possible to obtain a somewhat crude estimate of the temperature variation of the effective mass from the measured temperature dependence of Seebeck coefficient.

REFERENCES

1. F. Seitz, *Modern Theory of Solids*, McGraw-Hill, New York, 1940.
2. R.A. Hutner, E.S. Rittner and F.K. DuPré, 'Fermi Levels in Semiconductors', *Philips Research Reports* **5** (1950), 188–204.
3. J.R. Drabble and H.J. Goldsmid, *Thermal Conduction in Semiconductors*, Pergamon Press, Oxford, 1961.
4. P.G. Klemens, 'The thermal conductivity of dielectric solids at low temperatures', *Proceedings of the Royal Society*, **A208** (1951), 108–33.
5. M.S. Abrahams, R. Braunstein and F.D. Rosi, 'Thermal, electrical and optical properties of (In, Ga)As alloys', *Journal of Physics and Chemistry of Solids*, **10** (1959), 204–10.
6. H.J. Goldsmid and A.W. Penn, 'Boundary scattering of phonons in solid solutions', *Physics Letters*, **27A**, (1968), 523–4.
7. J.M. Ziman, *Electrons and Phonons*, Oxford University Press, Oxford, 1960.
8. C. Herring, 'Role of low energy phonons in thermal conduction', *Physical Review* **95** (1954), 954–65.
9. J. Callaway, 'Model for lattice thermal conductivity at low temperature', *Physical Review*, **113** (1959), 1046–51.
10. P.G. Klemens, 'The scattering of low frequency lattice waves by static imperfections', *Proceedings of the Physical Society* (*London*), **A68** (1955), 1113–28.
11. J.M. Ziman, 'The effect of free electrons on lattice conduction', *Philosophical Magazine* (*Ser 8*), **1** (1956), 191–8; **2** (1957), 292.
12. H.R. Meddins and J.E. Parrott, 'The thermal and thermoelectric properties of sintered germanium silicon alloys', *Journal of Physics C: Solid State Physics*, **9** (1976), 1263–76.
13. J.E. Parrott, 'The thermal conductivity of sintered semiconductor alloys', *Journal of Physics C*, **2** (1969), 147–51.
14. C.M. Bhandari and D.M. Rowe, 'Boundary scattering of phonons', *Journal of Physics C: Solid State Physics*, **11** (1978) 1787–94.

15. W. Shockley, *Electrons and Holes in Semiconductors*, Van Nostrand, New York, p. 178, 1950.
16. H.J. Goldsmid, *Application of Thermoelectricity*, Methuen, London, 1960.
17. See for example F.J. Blatt, *Solid State Physics*, Volume 4 (Ed.) F. Seitz and B. Turnbull. Academic Press, London, 1957.
18. N.K.S. Gaur, C.M. Bhandari and G.S. Verma, 'Temperature dependence of density of states effective mass and electronic and phonon contributions to thermal resistance of doped silicon germanium alloys at high temperatures', *Physical Review*, **144**(2) (1966), 628–41.
19. R.P. Chasmar and R.J. Stratton, 'The thermoelectric figure of merit and its relationship to thermoelectric generators', *Journal of Electronics and Control*, **7** (1959) 52–72.
20. R.W. Ure Jr, 'Practical limits to the thermoelectric figure of merit II', *Energy Conversion*, **12** (1972), 45–52.

4

Optimization of the Thermoelectric Figure of Merit—Theoretical Consideration

4.1 INTRODUCTION

THEORETICAL work relating to the figure of merit falls in two categories. The first is concerned with ascertaining whether or not a finite upper limit to the thermoelectric figure of merit exists, and if it does, whether this limit can be obtained in principle. In the second category, somewhat idealized models of a semiconductor are usually assumed and the conditions under which the figure of merit is maximized explored by substituting more or less realistic values for the various parameters.

Rittner[1] considered a very simple model in which both elements or legs of the thermocouple were assumed to be constructed from the same material, and in addition he assumed that:

1. Conduction takes place in a single band.
2. The energy surfaces are spherical.
3. Carrier mean free path is proportional to the rth power of the energy, where r is related to s (defined in Chapter 3) by $r = s + \frac{1}{2}$.
4. The lattice thermal conductivity is negligible.

A shortcoming of this model is that the Seebeck coefficient tends to approach infinity as the Fermi level diverges without limit from the band edge. This difficulty can be avoided by considering a second band or by making the additional assumption that, instead of diverging indefinitely from the band edge with decrease in carrier concentration, the Fermi level approaches the intrinsic value. The theoretical upper bound obtained with this model is:

$$Z_{max} = (r + 2 + E_g/2kT)^2/(r+2)T \qquad 4.1$$

where E_g is the intrinsic band gap. This equation can readily be obtained from

Equation 3.35, Chapter 3, by assuming that all heat is carried by electrons (holes) and substituting $\xi = -\frac{\xi_g}{2}$. This expression gives a value for the figure of merit which is an order of magnitude higher than that actually observed in the best materials.

Chasmar and Stratton[2] presented a detailed numerical evaluation of the theoretical figure of merit. The results were presented graphically with the figure of merit plotted against reduced Fermi potential and the so-called scattering parameter for various scattering mechanisms. They did not investigate the upper bound but rather obtained general results which could be applied to realistic physical situations. Littman and Davidson[3] in 1961 concluded from arguments of irreversible thermodynamics that the second law does not impose any upper bound on the figure of merit. However, it does not necessarily follow that Z can be increased indefinitely. Any investigations should be guided by the specific physical model under examination and the possibility of realizing it in practice.

A two-band model was adopted by Simon[4] and later by Rittner and Neumark[5] in order to obtain an upper bound to Z. Simon was unable to set a finite upper bound but did point out that such a bound could be shown to exist if certain limiting values and relationships between various physical parameters were taken into consideration.

4.2 A SIMPLE TWO-BAND MODEL

An insight into the conditions which favour a high upper bound to the thermoelectric figure of merit can be investigated by assuming a specific model. The general case of conduction in a two-band model with a multivalley structure has been discussed previously and expressions for the relevant transport coefficients derived. It should be noted that conduction in a single-band model with a single spherical valley can always be discussed as a special case of the generalized expression.

Considering a simple two-band conduction model, and assuming that the thermal conductivity λ results exclusively from free carriers we can write:

$$\lambda = \lambda_e = \left(\frac{k}{e}\right)^2 T \sum_{i=1}^{N} \mathscr{L}_i \sigma_i \tag{4.2}$$

the suffix i runs over N extrema in the conduction and valence bands.

For the total electrical conductivity we write:

$$\sum_i \sigma_i = N_e \sigma_{1e} + N_h \sigma_{1h} \tag{4.3}$$

where N_e and N_h are the number of equivalent extrema in the conduction and valence bands (non-degenerate) and σ_1 is the contribution to electrical conductivity from one extrema.

The Seebeck coefficient is written as:

$$\alpha = \frac{k}{e} \sum_{i=1}^{N} \frac{A_i \sigma_i}{\sigma} \tag{4.4}$$

where $A_i = \delta - \xi_i$ and δ_i is defined in *3.43*.
With the substitution $\sigma_e = N_e \sigma_{1e}$ and $\sigma_h = N_h \sigma_{1h}$:

$$\alpha = \frac{k}{e}\left(-\frac{A_e}{\left(1+\frac{\sigma_h}{\sigma_e}\right)} + \frac{A_h}{\left(1+\frac{\sigma_e}{\sigma_h}\right)}\right) \qquad 4.5$$

and the ratio of the electronic conductivity σ_e to the hole conductivity σ_h is given by:

$$\sigma_e/\sigma_h = \gamma^2 [F_{r_e}(\xi_e)/r_e!]/[F_{r_h}(\xi_h)/r_h!] \qquad 4.6$$

where $F_{r_1}(\xi_i)$ are Fermi integrals defined as:

$$F_{r_i}(\xi_i) = \int_0^\infty x^{r_i}[1+\exp(x-\xi_i)]^{-1}\,dx \qquad 4.7$$

and

$$\gamma = \left[\left(\frac{m_e^*}{m_h^*}\right)^{\frac{3}{2}}\left(\frac{\mu_{ce}}{\mu_{ch}}\right)\right]^{\frac{1}{2}} \qquad 4.8$$

Here μ_c represents the mobility in the low carrier concentration (classical) limit. m_e^* and m_h^* are the density of states effective masses for electrons and holes, respectively, and

$$m^* = N^{\frac{2}{3}}(m_1 m_2 m_3)^{\frac{1}{3}} \qquad 4.9$$

m_1, m_2, and m_3 being the masses along the principal directions.

Information on the upper bound to the figure of merit can be more easily extracted from this model if the assumptions are made that the electrons obey classical statistics and that equal scattering parameters are in operation for the electron and hole, i.e. $r_e = r_h$. In this case:

$$A_i = \pm[r+2-\xi_i] \qquad 4.10$$

$$\mathscr{L}_i = (r+2) \qquad 4.11$$

$$\frac{\sigma_e}{\sigma_h} = \gamma^2 e^{(\xi_g+2\xi)} \qquad 4.12$$

where $\xi_g = E_g/kT$ is again the reduced energy gap.

From these equations it is apparent that in this model $k/\sigma T$ is independent of ξ, ξ_g and γ and therefore a variation in these parameters will affect the figure of merit only through α. It is therefore sufficient to consider the dependence of α on these parameters. Equation 4.5 can now be written as:

$$\alpha = \frac{k}{e}(-M\tanh\xi' + \xi' - \ln\gamma) \qquad 4.13$$

where

$$\xi' = \xi + \xi_g/2 + \ln\gamma \qquad 4.14$$

and

$$M = r + 2 + \xi_g/2 \qquad 4.15$$

For a given γ, the Fermi level corresponding to the optimum Seebeck coefficient is given by:

For $\xi' > 0$,

$$\xi_{opt} = -\frac{\xi_g}{2} + \ln[(M)^{\frac{1}{2}} + (M-1)^{\frac{1}{2}}] - \ln\gamma \qquad 4.16$$

and $$\alpha_{opt} = -\frac{k}{e}\{[M(M-1)]^{\frac{1}{2}} - \ln[(M)^{\frac{1}{2}} + (M-1)^{\frac{1}{2}}] + \ln\gamma\} \quad 4.17$$

Evidently in this case an increase in γ decreases α.

For $\xi' < 0$, $$\xi_{opt} = -\frac{\xi_g}{2} - \ln[(M)^{\frac{1}{2}} + (M-1)^{\frac{1}{2}}] - \ln\gamma \quad 4.18$$

and $$\alpha_{opt} = \frac{k}{e}\{[M(M-1)]^{\frac{1}{2}} - \ln[(M)^{\frac{1}{2}} + (M-1)^{\frac{1}{2}}] - \ln\gamma\} \quad 4.19$$

Clearly a decrease in γ increases α.

It is evident from the above analysis that α increases as γ deviates from unity. This increase in α requires that the Fermi level should be adjusted to the value ξ_{opt}.

A detailed investigation[5] of the conditions under which the upper bound on the figure of merit can be raised reveals that the quantity γ is of central importance and can be related to the effective mass ratio m_h^*/m_e^* for particular scattering processes. In the case of lattice scattering it can be shown that:[5]

$$\gamma_L = \chi_L\left(\frac{m_h^*}{m_e^*}\right)^{\frac{1}{2}} \quad 4.20$$

where $$\chi_L = \left(\frac{N_e}{N_h}\right)^{\frac{5}{6}}\left(\frac{a_h}{a_e}\right)^{\frac{1}{2}}\left(\frac{\varepsilon_h}{\varepsilon_e}\right) \quad 4.21$$

and for ionized impurity scattering:

$$\gamma_I = \chi_I\left(\frac{m_e^*}{m_h^*}\right)^{\frac{1}{2}} \quad 4.22$$

where $$\chi_I = \left(\frac{N_e}{N_h}\right)^{\frac{1}{6}}\left(\frac{a_h}{a_e}\right)^{\frac{1}{4}} \quad 4.23$$

Here a is the mass anisotropy factor given by:

$$a = 3/[(m_2 m_3/m_1^2)^{\frac{1}{3}} + (m_1 m_3/m_2^2)^{\frac{1}{3}} + (m_1 m_2/m_3^2)^{\frac{1}{3}}] \quad 4.24$$

and ε refers to the deformation potentials.

The model predicts a large value for the upper bound on ZT provided that a high value for the ratio m_h^*/m_e^* and a small m_e^*/m_0 can be obtained. In addition, acoustic lattice scattering should be the dominant scattering process in the negative branch of the thermocouple and impurity scattering in the positive branch. The conduction band should also have a large number of extrema with a small number in the valence band of the negative branch. A large mass anisotropy for electrons, but not for holes, is another requirement for a large value of Z. The energy gap also influences the value of the upper bound. For a reduced energy gap of 4, $ZT_{max} = 8$, while for an energy gap of 16, $ZT_{max} \sim 70$, assuming a solubility limit of 10^{27} m^{-3}.

The results derived from such a model must be viewed with some scepticism, it being noted that these values to the upper bound are obtained assuming an unrealistically large ratio for m_h/m_e, absence of all lattice contributions to the thermal conductivity, and using classical statistics even at the solubility limit.

A general observation is that ZT_{max} should be large in materials with large energy gaps, and it would appear that such materials would be the best thermoelectric materials. Experimentally it is found that, although large energy gap materials are suitable for high temperature application, the highest figure of merit at around room temperature is found in small energy gap materials.

4.3 PRACTICAL LIMITS TO THE THERMOELECTRIC FIGURE OF MERIT

The thermoelectric figure of merit can be expressed in terms of a number of parameters, such as reduced Fermi energy, reduced energy gap, lattice thermal conductivity, carrier effective mass and the elastic constants. Change in composition and the addition of dopant controls the first two of these; consequently, if the other parameters are fixed, then the figure of merit can be optimized with respect to variations in the reduced Fermi energy and reduced energy gap. A simple case has been discussed earlier (in section 3.4.1), where using classical statistics Z has been optimized with respect to variations in the Seebeck coefficient. Further, using Fermi–Dirac statistics (section 3.4.2), the thermoelectric figure of merit has been expressed in terms of various generalized Fermi–Dirac functions and the results presented graphically. This enabled an estimate to be made of the maximum figure of merit.

Having discussed some highly idealized models and the theoretical upper bounds on the figure of merit, let us look at a more realistic situation, avoiding a many valley model for the moment and concentrating on two-band conduction in which we assume that the energy surfaces are spherical. Several of the parameters which determine the figure of merit depend upon the energy gap with a large ZT corresponding to a small E_g. However, a small E_g leads to conduction by minority carriers which will have a significant effect upon the Seebeck coefficient and electrical conductivity. The thermal conductivity will also change significantly due to the presence of bipolar terms.

The thermal conductivity λ is given by:

$$\lambda = \lambda_L + \lambda_e + \lambda_h + \lambda_b \qquad 4.25$$

where

$$\lambda_{(e,h)} = \left(\frac{k}{e}\right)^2 T \sigma_{e,h} \mathcal{L}_{e,h}(\xi_{e,h}) \qquad 4.26$$

$$\lambda_b = T\left(\frac{k}{e}\right)^2 \frac{\sigma_e \sigma_h}{\sigma}[\delta_e + \delta_h + \xi_g]^2 \qquad 4.27$$

(δ and $\mathcal{L}$ have been defined in Chapter 3.) The Seebeck coefficient is given by:

$$\alpha_{e,h} = \mp\frac{k}{e}\,[\delta(\xi_{e,h}) - \xi_{e,h}] \qquad 4.28$$

Assuming the same scattering parameter for both types of carrier, and defining reduced parameters σ' and α' where

$$\sigma' = \left(\frac{k}{e}\right)^2 \frac{T}{\lambda_L}\sigma \qquad 4.29$$

and
$$\alpha'_{e,h} = \mp\frac{e}{k}\alpha_{e,h} \tag{4.30}$$

one obtains

$$ZT = \frac{(\alpha'_h\sigma'_h - \alpha'_e\sigma'_e)^2}{(\sigma'_e + \sigma'_h)(1 + \sigma'_e\mathscr{L}_e + \sigma'_h\mathscr{L}_h) + \sigma'_e\sigma'_h(\delta_e + \delta_h + \xi_g)^2} \tag{4.31}$$

The reduced electrical conductivity for a single band is given by:

$$\sigma'_j = \frac{\beta F_r(\xi_j)}{r_j!} \tag{4.32}$$

where
$$\beta = \frac{2(2\pi k)^{\frac{3}{2}}k^2 A'}{eh^3} \tag{4.33}$$

and
$$A' = T^{\frac{5}{2}}m^{*\frac{3}{2}}\mu/\lambda_L \tag{4.34}$$

A'_h can be written as A'_e/b' where

$$b' = \left(\frac{m_e^*}{m_h^*}\right)^{\frac{3}{2}}\frac{\mu_e}{\mu_h} \tag{4.35}$$

The single-band case can readily be discussed in terms of these equations by taking $\sigma'_h = 0$ and neglecting the bipolar term.

Using these relations, Ure[6] has tried to estimate the practical upper limits to the figure of merit. His results indicate that the practical maximum is considerably smaller than the theoretical maximum proposed earlier. Experimental values for various parameters have been chosen from amongst a large number of possible thermoelectric materials.

Clearly, it is desirable to have a high value of b' and a low value of λ_L to achieve a higher upper limit of ZT. From amongst a large number of suitable materials the largest b' is around 7.6, and Ure takes $b' = 8$ in his calculations. In highly disordered crystals and amorphous materials λ_L is relatively low. However, their electrical properties are also poor and consequently these materials are not good choices for thermoelectric applications. One must therefore select materials for which the thermal conductivity is primarily determined by defect scattering (or boundary scattering in sintered material along with defect scattering) while mobility is still determined by phonon scattering. A comparison of room temperature λ_L values for various semiconducting materials gives lowest values in the range 0.4 to 1 W/Km.

The parameter A' has to be determined separately for various scattering mechanisms.

4.3.1 Acoustic Lattice Scattering

Except at very high impurity concentration, acoustic phonon scattering of charge carriers is the dominant scattering mechanism. In this case

$$A'_{ac} = \frac{(8\pi)^{\frac{1}{2}}e\hbar^4}{3k_0^{\frac{3}{2}}}\frac{C_{ll}T}{m^*\varepsilon_1^2\lambda_L} \tag{4.36}$$

with a large value of A' favouring a large figure of merit. Obviously one must find a lower limit for the effective mass and deformation potential coefficient ε_1 and an upper limit for the elastic constant C_{ll}. Although diamond has a very large elastic constant it is a very unlikely material for thermoelectric applications. Amongst others, silicon has a large value (1.7×10^{11} Nm^{-2}) and this value is shown as a suitable upper limit for C_{ll}. To obtain the deformation potential coefficient one can use experimental values of mobility, effective mass and elastic constants. The deformation potential coefficient has dimensions of energy and by convention values are quoted in electron volts (eV). A reasonable lower value selected by Ure is around 7 eV.

Effective mass has a very low value (0.014 m) in InSb. A mass this small is usually possible for low energy gaps of the order of 0.01 eV. In general, for a large number of semiconducting materials one can write $m^* = \text{constant}.E_g$, where the constant can be obtained from a plot of m^* and E_g for various materials.

Parameters ξ and ξ_g can now be varied to optimize ZT. The results of the calculation are presented in Table 4.1.

Table 4.1 *Maximum dimensionless figure of merit for two band semiconductor and scattering by acoustic mode phonons only* (after Ure, Reference 6). $\varepsilon_1 = 7$ eV.

λ_L(W/Km)	Reduced Fermi energy which optimizes ZT	ZT
1.0	−2.6	6.5
0.7	−2.9	7.4
0.4	−3.2	8.9

4.3.2 Optical Mode Scattering

In a number of semiconductors polar optical mode scattering is significant. One obtains A' from the expression of the mobility:[6]

$$A' = \frac{8h^2}{3e(2\pi k)^{\frac{1}{2}}} \frac{T^2 F(\Theta/T)}{\lambda_L(\varepsilon_\infty^{-1} - \varepsilon_0^{-1})} \qquad 4.37$$

with $\varepsilon_\infty^{-1} - \varepsilon_0^{-1} = 0.0072\, E_g$ (E_g in eV), and $F(x) = x^{-1}(e^x - 1)\chi(x)$; for $T > \frac{\Theta}{3}$, the function $\chi\left(\frac{\Theta}{T}\right)$ lies between 0.66 and 1. The constant in equation 4.38 is obtained from a plot of the experimental values of $(\varepsilon_\infty^{-1} - \varepsilon_0^{-1})$ versus E_g. Chosing Θ in the range 150–450 K, ZT has been optimized with respect to variations in ξ and ξ_g. At 300 K the calculated value of ZT is approximately 4.

An estimate can be made of ZT when both the scattering mechanisms are simultaneously present. Since ZT increases monotonically with μ, a rather crude estimate of ZT is obtained as:

$$(ZT)^{-1} = \sum_i (ZT)_i^{-1} \qquad 4.39$$

where $(ZT)_i$ is the value when only the ith scattering mechanism is present. For the values referred to above, ZT (for $\lambda_L = 0.4\,\mathrm{WK^{-1}\,m^{-1}}$) is approximately 1.9 at 300 K and 3.6 at 1200 K.

Although much lower than the maximum ZT obtained with more idealized models (like the one discussed in Section 4.2), it is still a factor of two larger than the observed values in best known materials. The possibility of finding or developing materials with these ZT values is not very high in the immediate future, but an improvement by 20 to 30 per cent might be possible. The material under discussion has electrical parameters similar to III-V compounds and silicon and germanium, while the parameters for thermal conductivity are closer to those of bismuth and antimony telluride. Thermoelectric semiconductors are usually very heavily doped in order to optimize the figure of merit. Consequently, the scattering of charge carriers by ionized impurities should also be taken into account.

4.3.3 Multivalley Semiconductor Systems

There are a number of semiconductors whose behaviour cannot be adequately described in terms of a simple spherical valley model. A so-called multivalley band structure is employed consisting of N_v equivalent valleys. The overall conductivity tensor due to the carriers in all the valleys is $\Sigma_i \sigma_{\alpha\beta}^{(i)}$ and the overall mobility tensor is the average of $\mu_{\alpha\beta}^{(i)}$ over the different valleys. Here $\sigma_{\alpha\beta}^{(i)}$, $(\mu_{\alpha\beta}^{(i)})$ refers to the conductivity (mobility) tensor component in the ith valley.

For non-spherical energy surfaces it is important to distinguish between an inertial effective mass m_i and a density of states mass m^*. In a single valley semiconductor and for scattering by acoustic phonons:

$$\mu \propto m_i^{-1} m^{*-\frac{3}{2}} \tag{4.40}$$

and

$$A' \propto m_i^{-1} \tag{4.41}$$

For a many valley semiconductor without intervalley scattering:

$$\mu \propto N_v m_i^{-1} m_t^{*-\frac{3}{2}} \tag{4.42}$$

As

$$m_t^{*\frac{3}{2}} = N_v m_s^{*\frac{3}{2}} \tag{4.43}$$

where m_s^* is the density of states mass for a single valley and m_t^* the overall density of states mass; the relationship implies that the mobility depends upon the inertial mass m_i and density of states effective mass corresponding to a single valley. In this case:

$$A' \propto N_v / m_i \tag{4.44}$$

The multivalley structure with a large number of equivalent valleys favours a large figure of merit, particularly when the inertial mass is small. On the other hand, intervalley scattering serves to increase the carrier scattering, resulting in a reduction in the figure of merit. However, it has been shown that a many valley model with a large number of valleys can account for the thermoelectric behaviour of Bi_2Te_3 and Si–Ge alloys, the former being a 'good' thermoelectric material at low temperature and the latter having application at high temperature.

In general it is necessary to consider a number of scattering mechanisms, as more than one mechanism may be operating in a certain situation. One simple method is to obtain ZT separately for the various mechanisms and then combine them to give a rather crude estimate of $(ZT)_{total}$ by adding their inverse values. A better method is to consider mixed scattering of two or more types of scattering and then obtain ZT. This procedure, although more rigorous in principle, leads to unavoidable complications which tend to offset its advantages.

The mechanism dominant in scattering charge carriers will determine ZT. As it is now well understood that the material for thermoelectric application should be partially degenerate,[8] it is quite obvious that a large concentration of ionized impurities will be an important scattering source for the carriers, and at moderate temperatures may be the single most significant factor in determining ZT. At high temperatures the situation may be different as the scattering by phonons, and in particular the intervalley scattering in a many-valley semiconductor, may be the most important mechanism in determining ZT.

It is worth pointing out that in bismuth telluride, a semiconducting compound having one of the highest figures of merit, a six- or twelve-valley model, is thought to apply both to the conduction and the valence hand. In addition, the inertial effective mass for carriers is about one tenth that of the free electron mass. Silicon and silicon-rich alloys of silicon and germanium also possess the multivalley structure.

Although the silicon–germanium alloys are not good thermoelectric materials at low and moderate temperatures, they are capable of operating at around 1300 K. The multivalley structure is probably responsible to an extent for a respectable figure of merit in n-type alloys in spite of the relatively high lattice thermal conductivity. The possibility of reducing the lattice thermal conductivity by the use of sintering techniques has been the theme of many papers during the last decade (see Chapter 5, references 7, 8, 9 and 10).

4.3.4 Mixed Scattering

In real systems, two or more carrier scattering mechanisms usually operate simultaneously. Initially the main scattering mechanisms must be identified and their contribution to the various transport properties determined. The contribution of each type of scattering varies strongly with temperature and with impurity concentration. In heavily doped materials a situation where only one mechanism is operative is rare, and normally at least two mechanisms need to be considered. There may be scattering by ions and acoustic phonons or scattering by ions and optical phonons. The first combination is operative in covalent semiconductors and the second in polar semiconductors including $A^{III}B^{V}$ compounds.

It is possible to consider mixed scattering for a many valley model, but the limitations imposed by our present level of understanding makes it a not very meaningful exercise. However, it is useful to consider the mixed scattering case for a single valley model, as various parameters frequently encountered in the analysis of data, such as Hall factor and Lorenz factor, depend upon the relative contributions of the scattering mechanisms. In the next section we have also attempted to give some of the results of ZT calculations using various models.

The general expression for the electrical conductivity of a semiconductor can be written as:

$$\sigma=\frac{8\pi e^2(2m^*kT)^{\frac{3}{2}}}{3m^*h^3}\int_0^{\infty}\frac{\partial f_0}{\partial\eta}\eta^{\frac{3}{2}}\tau(\eta)\mathrm{d}\eta \qquad 4.45$$

where η is the reduced carrier energy, E/kT.

Considering the mixed scattering case when acoustic phonons and impurity scattering act simultaneously with associated relaxation times τ_{ac} and τ_{imp}, the combined relaxation time is given by:

$$\frac{1}{\tau}=\sum_{i=1}^{i=2}\frac{1}{\tau_i} \qquad 4.46$$

where

$$\tau_{ac}=\tau_{oL}\eta^{-\frac{1}{2}} \qquad 4.47$$

and

$$\tau_{imp}=\tau_{oi}\eta^{\frac{3}{2}} \qquad 4.48$$

τ can thus be written as $\dfrac{\tau_{oL}\eta^{\frac{3}{2}}}{\eta^2+b^2}$

where $b^2=\tau_{oL}/\tau_{oi}$ is a measure of the relative strengths of ionized impurity scattering and lattice scattering. Substituting the expression for the combined relaxation time in the electrical conductivity equation gives:

$$\sigma=\frac{8\pi e^2(2m^*kT)^{\frac{3}{2}}}{3m^*h^3}\tau_L\phi_3(\xi,b) \qquad 4.49$$

where ϕ_n is a Fistul integral.
ϕ_n is a 'Fistul integral' and is given by

$$\phi_n=\int_0^{\infty}\frac{\eta^n\exp(\eta-\xi)\mathrm{d}\eta}{(\eta^2+b^2)[1+\exp(\eta-\xi)]^2} \qquad 4.50$$

where n is an integer (even or odd).

For fractional indices, the integral becomes

$$\phi_{K+\frac{1}{2}}=\int\frac{\eta^{K+\frac{1}{2}}\exp(\eta-\xi)\mathrm{d}\eta}{(\eta^2+b^2)^2[1+\exp(\eta-\xi)]^2} \qquad 4.51$$

where K is an integer.
Similarly the Hall coefficient R_H, Hall mobility μ_H, thermoelectric power α, Hall factor r_H, and the Lorenz number L can be expressed as:

$$R_H=\frac{3h^3}{8\pi e(2m^*kT)^{\frac{3}{2}}}\frac{\phi_{\frac{9}{2}}(\xi,b)}{\{\phi_3(\xi,b)\}^2} \qquad 4.52$$

$$\mu_H=\frac{e}{m^*}\tau_{oL}\frac{\phi_{\frac{9}{2}}(\xi,b)}{\phi_3(\xi,b)} \qquad 4.53$$

$$\alpha=-\frac{k}{e}\left\{\frac{\phi_4(\xi,b)}{\phi_3(\xi,b)}-\xi\right\} \qquad 4.54$$

$$r_H=\frac{\frac{3}{2}F_{\frac{1}{2}}(\xi)\phi_{\frac{9}{2}}(\xi,b)}{[\phi_3(\xi,b)]^2} \quad 4.55$$

$$L=\frac{k^2}{e^2}\mathscr{L}=\frac{k^2}{e^2}\frac{\phi_3\phi_5-\phi_4^2}{\phi_3^2} \quad 4.56$$

Values for ϕ_3, ϕ_4, ϕ_5 and $\phi_{\frac{9}{2}}$ have been tabulated in section 4.4.

In mixed scattering situations values for both b and ξ need to be known. Both parameters can be unequivocally determined from Hall mobility and Seebeck coefficient measurements as follows. In an impurity-free semiconductor when scattering is due to phonons only:

$$\mu_H\to\mu_{Ho}=\frac{\sqrt{\pi}}{2}\frac{e}{m^*}\tau_{oL} \quad 4.57$$

and

$$\mu_H=\frac{2}{\sqrt{\pi}}\mu_{HO}f_1(\xi,b) \quad 4.58$$

$$\alpha=-\frac{k}{e}f_2(\xi,b) \quad 4.59$$

where

$$f_1=\frac{\phi_{\frac{9}{2}}(\xi,b)}{\phi_3(\xi,b)} \quad 4.60$$

$$f_2=\frac{\phi_4(\xi,b)}{\phi_3(\xi,b)}-\xi \quad 4.61$$

f_1 and f_2 are plotted as functions of ξ and b in Figure 4.1. The experimental Hall mobility values and the mobility of a pure specimen give a value for f_1, while f_2 is obtained from the Seebeck coefficient. Values of ξ and b which satisfy both relationships are readily obtained from the figure. The Lorenz factor and Hall factor as functions of ξ and b are displayed in Figures 4.2 and 4.3. Appropriate expressions for the Hall factor and the Lorenz factor can readily be obtained from equations 4.55 and 4.56 when scattering takes place only by (a) phonons or (b) ionized impurities.

(a) *For phonon scattering only;* $b^2\ll\eta^2$

then

$$r_{H\,lattice}=\frac{3}{4}F_{\frac{1}{2}}(\xi)\frac{F_{-\frac{1}{2}}(\xi)}{F_0^2(\xi)} \quad 4.62$$

$$\mathscr{L}_{lattice}=\frac{3F_0(\xi)F_2(\xi)-4F_1^2(\xi)}{F_0^2(\xi)} \quad 4.63$$

In lightly doped material $\xi\ll0$

and

$$r_{H\,lattice}=\frac{3\pi}{8}=1.18 \quad 4.64$$

$$\mathscr{L}_{lattice}=2 \quad 4.65$$

In heavily doped material $\xi\gg0$

and

$$r_{H\,lattice}=1 \quad 4.66$$

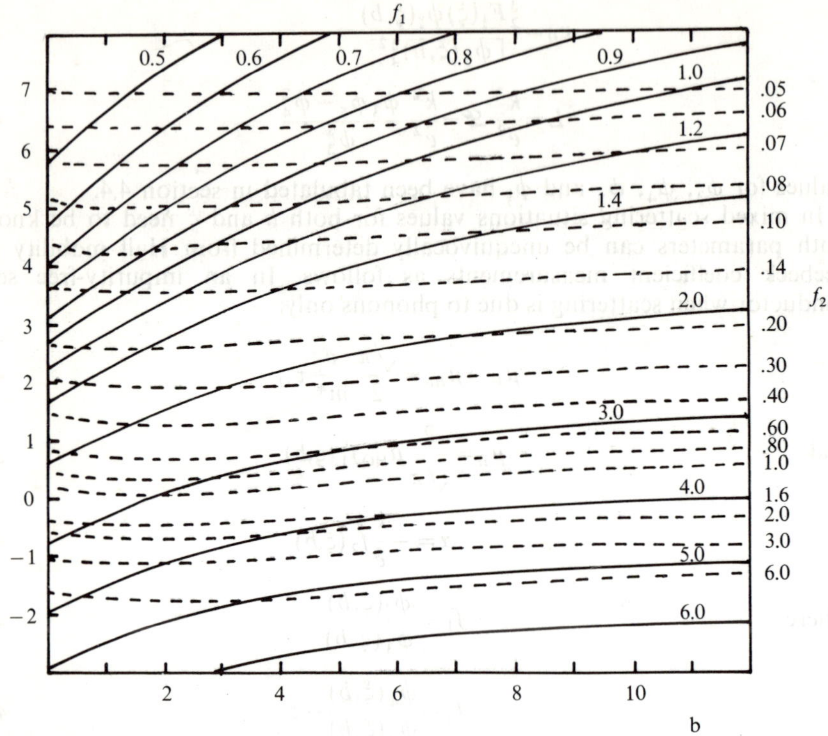

Figure 4.1 f_1 and f_2 as a function of ξ and b.

$$\mathcal{L}_{lattice} = \frac{\pi^2}{3}\left(1 - \frac{\pi^2}{3}\cdot\frac{1}{\xi^2}\right) \tag{4.67}$$

(b) *For ionized impurity scattering only;* $b^2 \gg \eta^2$

then

$$r_{H\ impurity} = \tfrac{3}{4} F_{\frac{1}{2}}(\xi) \frac{F_{\frac{7}{2}}(\xi)}{F_2^2(\xi)} \tag{4.68}$$

$$\mathcal{L}_{impurity} = \frac{15F_2(\xi)F_4(\xi) - 16F_3^2(\xi)}{9F_2^2(\xi)} \tag{4.69}$$

In lightly doped material $\xi \ll 0$

and

$$r_{H\ impurity} = \frac{315\pi}{512} = 1.93 \tag{4.70}$$

$$\mathcal{L}_{impurity} = 4 \tag{4.71}$$

In heavily doped material $\xi \gg 0$

and

$$r_{H\ impurity} = 1 \tag{4.72}$$

$$\mathcal{L}_{impurity} = \frac{1}{\xi^2 + \pi^2}\left[\xi^2 + \frac{10\pi^2}{3} - \frac{(\xi^3 + 2\pi^2)^2}{\xi^2 + \pi^2}\right] \tag{4.73}$$

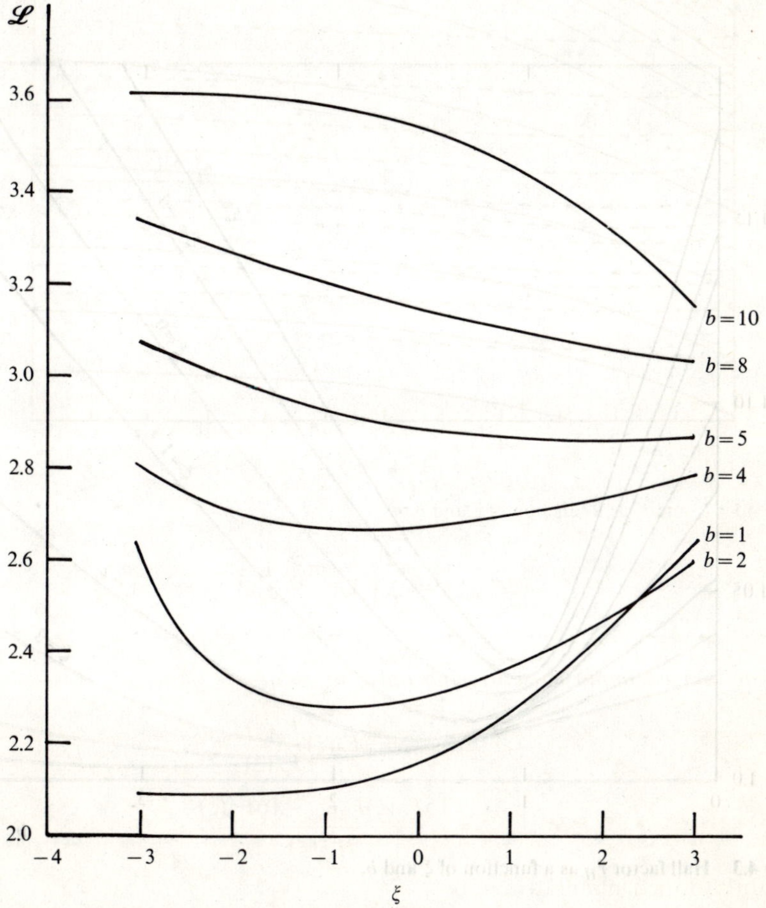

Figure 4.2 Lorenz factor $\mathscr{L}$ as a function of ξ and b.

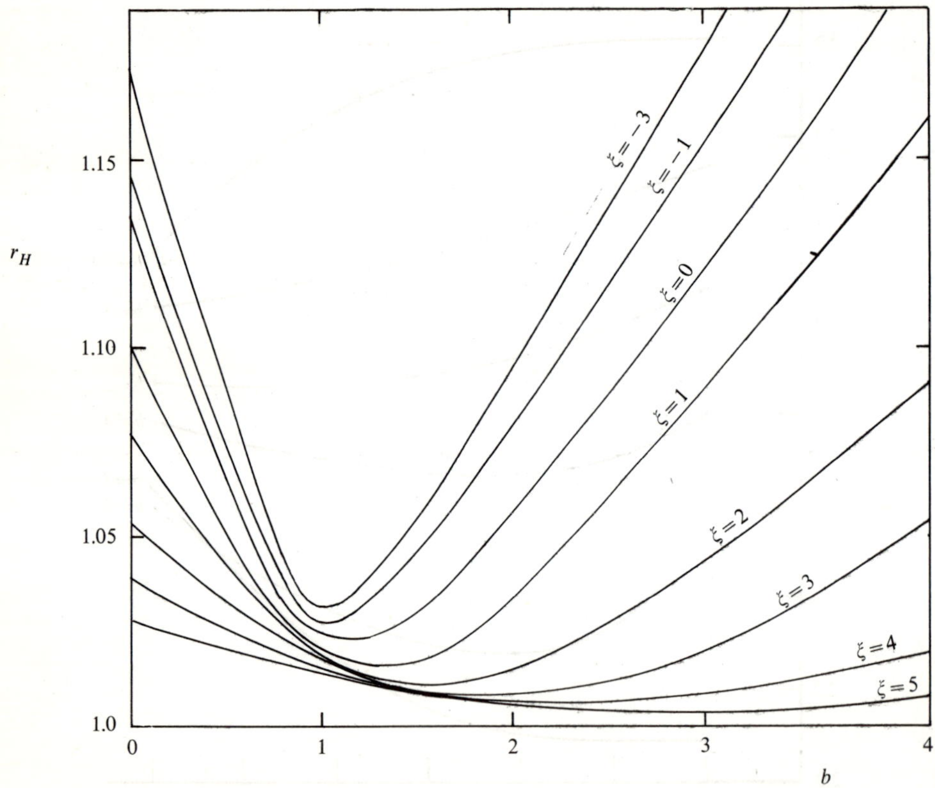

Figure 4.3 Hall factor r_H as a function of ξ and b.

4.4 TABLES OF INTEGRALS (MIXED SCATTERING)

A. Rowlands of the Department of Applied Physics, Electronics and Electrical Engineering, UWIST, computed the majority of the tabulated 'Fistul integrals', and his advice on their behaviour is appreciated.

When reading the following tables, the notation E − 3 for example means that the decimal place is moved three places to the left, and E + 2 indicates a movement of two places to the right.

Numerical values of integral $\phi_3(\xi, b)$

$\xi \downarrow$ \ $b \rightarrow$	−1	0	1	2	3	4
−3.0	3.2179995 E − 2	4.8587352 E − 2	3.2179995 E − 2	2.0751574 E − 2	1.4146380 E − 2	1.0122854 E − 2
−2.9	3.5506401 E − 2	5.3562776 E − 2	3.5506401 E − 2	2.2905621 E − 2	1.5617876 E − 2	1.1177102 E − 2
−2.8	3.9170149 E − 2	5.9032827 E − 2	3.9170149 E − 2	2.5280059 E − 2	1.7240598 E − 2	1.2339971 E − 2
−2.7	4.3204065 E − 2	6.5043562 E − 2	4.3204065 E − 2	2.7896755 E − 2	1.9029692 E − 2	1.3622402 E − 2
−2.6	4.7643877 E − 2	7.1644692 E − 2	4.7643877 E − 2	3.0779596 E − 2	2.1001739 E − 2	1.5036383 E − 2
−2.5	5.2528416 E − 2	7.8889734 E − 2	5.2528416 E − 2	3.3954653 E − 2	2.3174874 E − 2	1.6595041 E − 2
−2.4	5.7899805 E − 2	8.6836152 E − 2	5.7899805 E − 2	3.7450343 E − 2	2.5568904 E − 2	1.8312735 E − 2
−2.3	6.3803662 E − 2	9.5545465 E − 2	6.3803662 E − 2	4.1297593 E − 2	2.8205446 E − 2	2.0205155 E − 2
−2.2	7.0289280 E − 2	1.0508332 E − 1	7.0289280 E − 2	4.5530016 E − 2	3.1108057 E − 2	2.2289431 E − 2
−2.1	7.7409806 E − 2	1.1551952 E − 1	7.7409806 E − 2	5.0184079 E − 2	3.4302378 E − 2	2.4584236 E − 2
−2.0	8.5222400 E − 2	1.2692801 E − 1	8.5222400 E − 2	5.5299276 E − 2	3.7816269 E − 2	2.7109904 E − 2
−1.9	9.3788370 E − 2	1.3938676 E − 1	9.3788370 E − 2	6.0918296 E − 2	4.1679957 E − 2	2.9888542 E − 2
−1.8	1.0317328 E − 1	1.5297761 E − 1	1.0317328 E − 1	6.7087169 E − 2	4.5926180 E − 2	3.2944150 E − 2
−1.7	1.1344700 E − 1	1.6778603 E − 1	1.1344700 E − 1	7.3855431 E − 2	5.0590320 E − 2	3.6302736 E − 2
−1.6	1.2468376 E − 1	1.8390074 E − 1	1.2468376 E − 1	8.1276235 E − 2	5.5710542 E − 2	3.9992437 E − 2
−1.5	1.3696206 E − 1	2.0141328 E − 1	1.3696206 E − 1	8.9406463 E − 2	6.1327924 E − 2	4.4043625 E − 2
−1.4	1.5036460 E − 1	2.2041741 E − 1	1.5036460 E − 1	9.8306799 E − 2	6.7486555 E − 2	4.8489026 E − 2
−1.3	1.6497809 E − 1	2.4100846 E − 1	1.6497809 E − 1	1.0804177 E − 1	7.4233642 E − 2	5.3363810 E − 2
−1.2	1.8089295 E − 1	2.6328247 E − 1	1.8089295 E − 1	1.1867975 E − 1	8.1619581 E − 2	5.8705685 E − 2
−1.1	1.9820295 E − 1	2.8733533 E − 1	1.9820295 E − 1	1.3029287 E − 1	8.9698007 E − 2	6.4554971 E − 2
−1.0	2.1700476 E − 1	3.1326169 E − 1	2.1700476 E − 1	1.4295698 E − 1	9.8525809 E − 2	7.0954659 E − 2
−0.9	2.3739729 E − 1	3.4115388 E − 1	2.3739729 E − 1	1.5675138 E − 1	1.0816311 E − 1	7.7950447 E − 2
−0.8	2.5948107 E − 1	3.7110067 E − 1	2.5948107 E − 1	1.7175864 E − 1	1.1867322 E − 1	8.5590750 E − 2
−0.7	2.8335734 E − 1	4.0318605 E − 1	2.8335734 E − 1	1.8806425 E − 1	1.3012249 E − 1	9.3926688 E − 2
−0.6	3.0912719 E − 1	4.3748795 E − 1	3.0912719 E − 1	2.0575617 E − 1	1.4258022 E − 1	1.0301203 E − 1
−0.5	3.3689045 E − 1	4.7407699 E − 1	3.3689045 E − 1	2.2492444 E − 1	1.5611838 E − 1	1.1290312 E − 1
−0.4	3.6674463 E − 1	5.1301525 E − 1	3.6674463 E − 1	2.4566047 E − 1	1.7081137 E − 1	1.2365876 E − 1
−0.3	3.9878365 E − 1	5.5435525 E − 1	3.9878365 E − 1	2.6805647 E − 1	1.8673570 E − 1	1.3534000 E − 1
−0.2	4.3309664 E − 1	5.9813887 E − 1	4.3309664 E − 1	2.9220469 E − 1	2.0396956 E − 1	1.4801003 E − 1
−0.1	4.6976664 E − 1	6.4439666 E − 1	4.6976664 E − 1	3.1819658 E − 1	2.2259243 E − 1	1.6173382 E − 1
0.0	5.0886935 E − 1	6.9314718 E − 1	5.0886935 E − 1	3.4612195 E − 1	2.4268453 E − 1	1.7657793 E − 1
0.1	5.5047189 E − 1	7.4439668 E − 1	5.5047189 E − 1	3.7606802 E − 1	2.6432624 E − 1	1.9261010 E − 1
0.2	5.9463173 E − 1	7.9813889 E − 1	5.9463173 E − 1	4.0811851 E − 1	2.8759759 E − 1	2.0989893 E − 1
0.3	6.4139561 E − 1	8.5435524 E − 1	6.4139561 E − 1	4.4235268 E − 1	3.1257751 E − 1	2.2851349 E − 1
0.4	6.9079872 E − 1	9.1301525 E − 1	6.9079872 E − 1	4.7884434 E − 1	3.3934326 E − 1	2.4852283 E − 1
0.5	7.4286401 E − 1	9.7407698 E − 1	7.4286401 E − 1	5.1766107 E − 1	3.6796969 E − 1	2.6999558 E − 1
0.6	7.9760166 E − 1	1.0374879	7.9760166 E − 1	5.5886318 E − 1	3.9852860 E − 1	2.9299942 E − 1
0.7	8.5500896 E − 1	1.1031860	8.5500896 E − 1	6.0250304 E − 1	4.3108803 E − 1	3.1760065 E − 1
0.8	9.1507016 E − 1	1.1711007	9.1507016 E − 1	6.4862440 E − 1	4.6571163 E − 1	3.4386367 E − 1
0.9	9.7775672 E − 1	1.2411539	9.7775672 E − 1	6.9726181 E − 1	5.0245806 E − 1	3.7185050 E − 1
1.0	1.0430278	1.3132617	1.0430278	7.4844014 E − 1	5.4138038 E − 1	4.0162031 E − 1
1.1	1.1108308	1.3873353	1.1108308	8.0217431 E − 1	5.8252559 E − 1	4.3322897 E − 1
1.2	1.1811022	1.4632825	1.1811022	8.5846909 E − 1	6.2593413 E − 1	4.6672864 E − 1
1.3	1.2537688	1.5410085	1.2537688	9.1731910 E − 1	6.7163959 E − 1	5.0216736 E − 1
1.4	1.3287483	1.6204174	1.3287483	9.7870895 E − 1	7.1966834 E − 1	5.3958867 E − 1
1.5	1.4059512	1.7014132	1.4059512	1.0426134	7.7003940 E − 1	5.7903140 E − 1
1.6	1.4852814	1.7839007	1.4852814	1.1089980	8.2276430 E − 1	6.2052932 E − 1
1.7	1.5666379	1.8677860	1.5666379	1.1778192	8.7784696 E − 1	6.6411096 E − 1
1.8	1.6499160	1.9529776	1.6499160	1.2490251	9.3528395 E − 1	7.0979948 E − 1
1.9	1.7350083	2.0393868	1.7350083	1.3225565	9.9506442 E − 1	7.5761252 E − 1
2.0	1.8218064	2.1269280	1.8218064	1.3983472	1.0571704	8.0756220 E − 1
2.1	1.9102015	2.2155195	1.9102015	1.4763246	1.1215772	8.5965506 E − 1
2.2	2.0000856	2.3050833	2.0000856	1.5564114	1.1882534	9.1389218 E − 1
2.3	2.0913522	2.3955455	2.0913522	1.6385255	1.2571616	9.7026921 E − 1

Numerical values of integral $\phi_3(\xi, b)$

$b\rightarrow$	5	6	7	8	9	10
−3.0	7.5348350 E−3	5.7911381 E−3	4.5699091 E−3	3.6863469 E−3	3.0292493 E−3	2.5289325 E−3
−2.9	8.3201525 E−3	6.3950294 E−3	5.0466251 E−3	4.0709945 E−3	3.3453956 E−3	2.7929037 E−3
−2.8	9.1865140 E−3	7.0613093 E−3	5.5726287 E−3	4.4954330 E−3	3.6942601 E−3	3.0842023 E−3
−2.7	1.0142110 E−2	7.7962980 E−3	6.1529209 E−3	4.9637047 E−3	4.0791696 E−3	3.4056087 E−3
−2.6	1.1195923 E−2	8.6069289 E−3	6.7929902 E−3	5.4802465 E−3	4.5037764 E−3	3.7601762 E−3
−2.5	1.2357795 E−2	9.5008034 E−3	7.4988555 E−3	6.0499260 E−3	4.9720879 E−3	4.1512549 E−3
−2.4	1.3638500 E−2	1.0486248 E−2	8.2771126 E−3	6.6780787 E−3	5.4884976 E−3	4.5825188 E−3
−2.3	1.5049824 E−2	1.1572376 E−2	9.1349837 E−3	7.3705483 E−3	6.0578190 E−3	5.0579933 E−3
−2.2	1.6604645 E−2	1.2769152 E−2	1.0080369 E−2	8.1337300 E−3	6.6853207 E−3	5.5820858 E−3
−2.1	1.8317019 E−2	1.4087460 E−2	1.1121904 E−2	8.9746153 E−3	7.3767646 E−3	6.1596173 E−3
−2.0	2.0202268 E−2	1.5539175 E−2	1.2269013 E−2	9.9008404 E−3	8.1384458 E−3	6.7958558 E−3
−1.9	2.2277077 E−2	1.7137239 E−2	1.3531975 E−2	1.0920735 E−2	8.9772338 E−3	7.4965520 E−3
−1.8	2.4559582 E−2	1.8895735 E−2	1.4921983 E−2	1.2043377 E−2	9.9006166 E−3	8.2679761 E−3
−1.7	2.7069473 E−2	2.0829967 E−2	1.6451212 E−2	1.3278642 E−2	1.0916746 E−2	9.1169562 E−3
−1.6	2.9828087 E−2	2.2956542 E−2	1.8132881 E−2	1.4637263 E−2	1.2034484 E−2	1.0050918 E−2
−1.5	3.2858507 E−2	2.5293448 E−2	1.9981324 E−2	1.6130884 E−2	1.3263450 E−2	1.1077925 E−2
−1.4	3.6185652 E−2	2.7860134 E−2	2.2012055 E−2	1.7772119 E−2	1.4614071 E−2	1.2206721 E−2
−1.3	3.9836374 E−2	3.0677590 E−2	2.4241835 E−2	1.9574607 E−2	1.6097626 E−2	1.3446770 E−2
−1.2	4.3839537 E−2	3.3768419 E−2	2.6688737 E−2	2.1553064 E−2	1.7726297 E−2	1.4808298 E−2
−1.1	4.8226094 E−2	3.7156909 E−2	2.9372206 E−2	2.3723344 E−2	1.9513215 E−2	1.6302334 E−2
−1.0	5.3029156 E−2	4.0869096 E−2	3.2313121 E−2	2.6102485 E−2	2.1472502 E−2	1.7940746 E−2
−0.9	5.8284045 E−2	4.4932818 E−2	3.5533839 E−2	2.8708756 E−2	2.3619318 E−2	1.9736284 E−2
−0.8	6.4028329 E−2	4.9377764 E−2	3.9058249 E−2	3.1561699 E−2	2.5969891 E−2	2.1702608 E−2
−0.7	7.0301845 E−2	5.4235503 E−2	4.2911801 E−2	3.4682168 E−2	2.8541559 E−2	2.3854321 E−2
−0.6	7.7146694 E−2	5.9539501 E−2	4.7121536 E−2	3.8092353 E−2	3.1352793 E−2	2.6206996 E−2
−0.5	8.4607223 E−2	6.5325134 E−2	5.1716103 E−2	4.1815803 E−2	3.4423217 E−2	2.8777195 E−2
−0.4	9.2729967 E−2	7.1629659 E−2	5.6725756 E−2	4.5877433 E−2	3.7773628 E−2	3.1582486 E−2
−0.3	1.0156358 E−1	7.8492189 E−2	6.2182343 E−2	5.0303526 E−2	4.1425993 E−2	3.4641451 E−2
−0.2	1.1115871 E−1	8.5953633 E−2	6.8119280 E−2	5.5121721 E−2	4.5403454 E−2	3.7973687 E−2
−0.1	1.2156788 E−1	9.4056620 E−2	7.4571506 E−2	6.0360987 E−2	4.9730308 E−2	4.1599801 E−2
0.0	1.3284533 E−1	1.0284540 E−1	8.1575422 E−2	6.6051583 E−2	5.4431989 E−2	4.5541393 E−2
0.1	1.4504679 E−1	1.1236571 E−1	8.9168813 E−2	7.2225013 E−2	5.9535028 E−2	4.9821036 E−2
0.2	1.5822924 E−1	1.2266464 E−1	9.7390748 E−2	7.8913954 E−2	6.5067014 E−2	5.4462241 E−2
0.3	1.7245072 E−1	1.3379048 E−1	1.0628147 E−1	8.6152185 E−2	7.1056535 E−2	5.9489419 E−2
0.4	1.8776998 E−1	1.4579248 E−1	1.1588229 E−1	9.3974492 E−2	7.7533114 E−2	6.4927833 E−2
0.5	2.0424619 E−1	1.5872071 E−1	1.2623538 E−1	1.0241656 E−1	8.4527134 E−2	7.0803539 E−2
0.6	2.2193865 E−1	1.7262579 E−1	1.3738369 E−1	1.1151489 E−1	9.2069750 E−2	7.7143324 E−2
0.7	2.4090641 E−1	1.8755867 E−1	1.4937073 E−1	1.2130661 E−1	1.0019280 E−1	8.3974633 E−2
0.8	2.6120795 E−1	2.0357037 E−1	1.6224042 E−1	1.3182941 E−1	1.0892871 E−1	9.1325492 E−2
0.9	2.8290081 E−1	2.2071175 E−1	1.7603688 E−1	1.4312138 E−1	1.1831037 E−1	9.9224433 E−2
1.0	3.0604123 E−1	2.3903321 E−1	1.9080424 E−1	1.5522084 E−1	1.2837105 E−1	1.0770039 E−1
1.1	3.3068384 E−1	2.5858446 E−1	2.0658644 E−1	1.6816621 E−1	1.3914424 E−1	1.1678262 E−1
1.2	3.5688126 E−1	2.7941425 E−1	2.2342707 E−1	1.8199589 E−1	1.5066354 E−1	1.2650062 E−1
1.3	3.8468381 E−1	3.0157011 E−1	2.4136910 E−1	1.9674803 E−1	1.6296275 E−1	1.3688401 E−1
1.4	4.1413921 E−1	3.2509810 E−1	2.6045478 E−1	2.1246047 E−1	1.7607527 E−1	1.4796246 E−1
1.5	4.4529225 E−1	3.5004258 E−1	2.8072535 E−1	2.2917050 E−1	1.9003447 E−1	1.5976557 E−1
1.6	4.7818460 E−1	3.7644599 E−1	3.0222095 E−1	2.4691481 E−1	2.0487333 E−1	1.7232283 E−1
1.7	5.1285451 E−1	4.0434863 E−1	3.2498037 E−1	2.6572926 E−1	2.2062437 E−1	1.8566344 E−1
1.8	5.4933662 E−1	4.3378847 E−1	3.4904096 E−1	2.8564882 E−1	2.3731956 E−1	1.9981631 E−1
1.9	5.8766184 E−1	4.6480101 E−1	3.7443840 E−1	3.0670737 E−1	2.5499018 E−1	2.1480991 E−1
2.0	6.2785712 E−1	4.9741906 E−1	4.0120665 E−1	3.2893766 E−1	2.7366678 E−1	2.3067223 E−1
2.1	6.6994545 E−1	5.3167269 E−1	4.2937774 E−1	3.5237114 E−1	2.9337901 E−1	2.4743068 E−1
2.2	7.1394568 E−1	5.6758904 E−1	4.5898170 E−1	3.7703788 E−1	3.1415560 E−1	2.6511202 E−1
2.3	7.5987258 E−1	6.0519229 E−1	4.9004648 E−1	4.0296649 E−1	3.3602422 E−1	2.8374228 E−1

Numerical values of integral $\phi_3(\xi, b)$

$\xi \downarrow$ $b \rightarrow$	−1	0	1	2	3	4
2.4	2.1838977	2.4868362	2.1838977	1.7225814	1.3282588	1.0287765
2.5	2.2776211	2.5788897	2.2776211	1.8084909	1.4014965	1.0893995
2.6	2.3724255	2.6716447	2.3724255	1.8961638	1.4768218	1.1521185
2.7	2.4682178	2.7650436	2.4682178	1.9855085	1.5541774	1.2169093
2.8	2.5649095	2.8590328	2.5649095	2.0764334	1.6335024	1.2837434
2.9	2.6624168	2.9535628	2.6624168	2.1688467	1.7147326	1.3525880
3.0	2.7606606	3.0485874	2.7606606	2.2626574	1.7978015	1.4234069
3.1	2.8595668	3.1440639	2.8595668	2.3577760	1.8826400	1.4961600
3.2	2.9590663	3.2399532	2.9590663	2.4541147	1.9691778	1.5708033
3.3	3.0590946	3.3362192	3.0590946	2.5515877	2.0573431	1.6472942
3.4	3.1595922	3.4328284	3.1595922	2.6501121	2.1470635	1.7255810
3.5	3.2605041	3.5297503	3.2605041	2.7496076	2.2382664	1.8056146
3.6	3.3617798	3.6269570	3.3617798	2.8499971	2.3308791	1.8873425
3.7	3.4633730	3.7244228	3.4633730	2.9512065	2.4248294	1.9707110
3.8	3.5652415	3.8221242	3.5652415	3.0531654	2.5200459	2.0556651
3.9	3.6673467	3.9200397	3.6673467	3.1558068	2.6164581	2.1421485
4.0	3.7696539	4.0181499	3.7696539	3.2590672	2.7139965	2.2301057

Numerical values of integral $\phi_4(\xi, b)$

$\xi \downarrow$ $b \rightarrow$	−1	0	1	2	3	4
−3.0	8.0000307 E−2	9.8361440 E−2	8.0000307 E−2	5.8945913 E−2	4.3703900 E−2	3.3107479 E-
−2.9	8.8331010 E−2	1.0856853 E−1	8.8331010 E−2	6.5098507 E−2	4.8271487 E−2	3.6570317 E-
−2.8	9.7519839 E−2	1.1981952 E−1	9.7519839 E−2	7.1887980 E−2	5.3313169 E−2	4.0393174 E-
−2.7	1.0765317 E−1	1.3221774 E−1	1.0765317 E−1	7.9379112 E−2	5.8877447 E−2	4.4613006 E-
−2.6	1.1882567 E−1	1.4587629 E−1	1.1882567 E−1	8.7643029 E−2	6.5017634 E−2	4.9270468 E-
−2.5	1.3114093 E−1	1.6091854 E−1	1.3114093 E−1	9.6757770 E−2	7.1792297 E−2	5.4410253 E-
−2.4	1.4471231 E−1	1.7747894 E−1	1.4471231 E−1	1.0680890 E−1	7.9265741 E−2	6.0081473 E-
−2.3	1.5966363 E−1	1.9570385 E−1	1.5966363 E−1	1.1789017 E−1	8.7508519 E−2	6.6338063 E-
−2.2	1.7613005 E−1	2.1575235 E−1	1.7613005 E−1	1.3010419 E−1	9.6597993 E−2	7.3239218 E-
−2.1	1.9425889 E−1	2.3779706 E−1	1.9425889 E−1	1.4356318 E−1	1.0661891 E−1	8.0849862 E-
−2.0	2.1421053 E−1	2.6202497 E−1	2.1421053 E−1	1.5838974 E−1	1.1766405 E−1	8.9241150 E-
−1.9	2.3615931 E−1	2.8863827 E−1	2.3615931 E−1	1.7471764 E−1	1.2983488 E−1	9.8491003 E-
−1.8	2.6029442 E−1	3.1785491 E−1	2.6029442 E−1	1.9269270 E−1	1.4324224 E−1	1.0868468 E-
−1.7	2.8682087 E−1	3.4991048 E−1	2.8682087 E−1	2.1247365 E−1	1.5800711 E−1	1.1991536 E-
−1.6	3.1596034 E−1	3.8505663 E−1	3.1596034 E−1	2.3423298 E−1	1.7426137 E−1	1.3228482 E-
−1.5	3.4795210 E−1	4.2356396 E−1	3.4795210 E−1	2.5815793 E−1	1.9214856 E−1	1.4590405 E-
−1.4	3.8305379 E−1	4.6572138 E−1	3.8305379 E−1	2.8445132 E−1	2.1182477 E−1	1.6089395 E
−1.3	4.2154226 E−1	5.1183672 E−1	4.2154226 E−1	3.1333252 E−1	2.3345942 E−1	1.7738610 E
−1.2	4.6371418 E−1	5.6223696 E−1	4.6371418 E−1	3.4503829 E−1	2.5723611 E−1	1.9552343 E
−1.1	5.0988663 E−1	6.1726830 E−1	5.0988663 E−1	3.7982365 E−1	2.8335350 E−1	2.1546101 E
−1.0	5.6039757 E−1	6.7729599 E−1	5.6039757 E−1	4.1796266 E−1	3.1202609 E−1	2.3736676 E
−0.9	6.1560601 E−1	7.4270403 E−1	6.1560601 E−1	4.5974916 E−1	3.4348511 E−1	2.6142230 E
−0.8	6.7589218 E−1	8.1389453 E−1	6.7589218 E−1	5.0549734 E−1	3.7797919 E−1	2.8782360 E
−0.7	7.4165732 E−1	8.9128689 E−1	7.4165732 E−1	5.5554230 E−1	4.1577515 E−1	3.1678175 E
−0.6	8.1332328 E−1	9.7531672 E−1	8.1332328 E−1	6.1024035 E−1	4.5715862 E−1	3.4852367 E
−0.5	8.9133191 E−1	1.0664346	8.9133191 E−1	6.6996924 E−1	5.0243460 E−1	3.8329275 E
−0.4	9.7614410 E−1	1.1651042	9.7614410 E−1	7.3512814 E−1	5.5192791 E−1	4.2134943 E
−0.3	1.0682385	1.2718008	1.0682385	8.0613742 E−1	6.0598350 E−1	4.6297175 E
−0.2	1.1681102	1.3870092	1.1681102	8.8343822 E−1	6.6496663 E−1	5.0845575 E
−0.1	1.2762685	1.5112213	1.2762685	9.6749177 E−1	7.2926290 E−1	5.5811585 E
0.0	1.3932353	1.6449341	1.3932353	1.0587784	7.9927805 E−1	6.1228504 E

Numerical values of integral $\phi_3(\xi, b)$

$b \rightarrow$	5	6	7	8	9	10
2.4	8.0773675 E − 1	6.4450359 E − 1	5.2259782 E − 1	4.3018400 E − 1	3.5901148 E − 1	3.0334674 E − 1
2.5	8.5754469 E − 1	6.8554096 E − 1	5.5665920 E − 1	4.5871584 E − 1	3.8314277 E − 1	3.2394979 E − 1
2.6	9.0929883 E − 1	7.2831932 E − 1	5.9225181 E − 1	4.8858574 E − 1	4.0844229 E − 1	3.4557496 E − 1
2.7	9.6299760 E − 1	7.7285046 E − 1	6.2939410 E − 1	5.1981539 E − 1	4.3493272 E − 1	3.6824464 E − 1
2.8	1.0186349	8.1914260 E − 1	6.6810327 E − 1	5.5242557 E − 1	4.6263606 E − 1	3.9198077 E − 1
2.9	1.0762034	8.6720266 E − 1	7.0839325 E − 1	5.8643457 E − 1	4.9157247 E − 1	4.1680384 E − 1
3.0	1.1356884	9.1703188 E − 1	7.5027507 E − 1	6.2185846 E − 1	5.2176034 E − 1	4.4273303 E − 1
3.1	1.1970733	9.6863021 E − 1	7.9375830 E − 1	6.5871213 E − 1	5.5321724 E − 1	4.6978688 E − 1
3.2	1.2603410	1.0219943	8.3884981 E − 1	6.9700839 E − 1	5.8595906 E − 1	4.9798265 E − 1
3.3	1.3254663	1.0771181	8.8555415 E − 1	7.3675816 E − 1	6.2000025 E − 1	5.2733645 E − 1
3.4	1.3924242	1.1339926	9.3387354 E − 1	7.7797050 E − 1	6.5535376 E − 1	5.5786323 E − 1
3.5	1.4611865	1.1926065	9.8380796 E − 1	8.2065263 E − 1	6.9203109 E − 1	5.8957679 E − 1
3.6	1.5317223	1.2529456	1.0353552	8.6480995 E − 1	7.3004225 E − 1	6.2248975 E − 1
3.7	1.6039984	1.3149936	1.0885110	9.1044606 E − 1	7.6939574 E − 1	6.5661350 E − 1
3.8	1.6779795	1.3787321	1.1432688	9.5756311 E − 1	8.1009868 E − 1	6.9195829 E − 1
3.9	1.7536286	1.4441401	1.1996203	1.0061601	8.5215669 E − 1	7.2853320 E − 1
4.0	1.8309068	1.5111952	1.2575552	1.0562367	8.9557401 E − 1	7.6634611 E − 1

Numerical values of integral $\phi_4(\xi, b)$

$b \rightarrow$	5	6	7	8	9	10
-3.0	2.5664975 E − 2	2.0326393 E − 2	1.6409173 E − 2	1.3472062 E − 2	1.1225743 E − 2	9.4767060 E − 3
-2.9	2.8350735 E − 2	2.2454225 E − 2	1.8127364 E − 2	1.4882968 E − 2	1.2401558 E − 2	1.0469429 E − 2
-2.8	3.1316021 E − 2	2.4803677 E − 2	2.0024601 E − 2	1.6440957 E − 2	1.3699985 E − 2	1.1565694 E − 2
-2.7	3.4589591 E − 2	2.7397585 E − 2	2.2119357 E − 2	1.8161213 E − 2	1.5133689 E − 2	1.2776203 E − 2
-2.6	3.8203096 E − 2	3.0261090 E − 2	2.4431968 E − 2	2.0060460 E − 2	1.6716620 E − 2	1.4112741 E − 2
-2.5	4.2191347 E − 2	3.3421850 E − 2	2.6984814 E − 2	2.2157103 E − 2	1.8464136 E − 2	1.5588288 E − 2
-2.4	4.6592618 E − 2	3.6910288 E − 2	2.9802518 E − 2	2.4471397 E − 2	2.0393138 E − 2	1.7217127 E − 2
-2.3	5.1448965 E − 2	4.0759846 E − 2	3.2912157 E − 2	2.7025618 E − 2	2.2522217 E − 2	1.9014973 E − 2
-2.2	5.6806576 E − 2	4.5007269 E − 2	3.6343487 E − 2	2.9844255 E − 2	2.4871817 E − 2	2.0999106 E − 2
-2.1	6.2716147 E − 2	4.9692907 E − 2	4.0129197 E − 2	3.2954214 E − 2	2.7464401 E − 2	2.3188520 E − 2
-2.0	6.9233281 E − 2	5.4861039 E − 2	4.4305175 E − 2	3.6385046 E − 2	3.0324645 E − 2	2.5604082 E − 2
-1.9	7.6418922 E − 2	6.0560226 E − 2	4.8910791 E − 2	4.0169178 E − 2	3.3479638 E − 2	2.8268701 E − 2
-1.8	8.4339812 E − 2	6.6843683 E − 2	5.3989215 E − 2	4.4342177 E − 2	3.6959093 E − 2	3.1207515 E − 2
-1.7	9.3068982 E − 2	7.3769682 E − 2	5.9587735 E − 2	4.8943020 E − 2	4.0795586 E − 2	3.4448088 E − 2
-1.6	1.0268627 E − 1	8.1401972 E − 2	6.5758121 E − 2	5.4014394 E − 2	4.5024803 E − 2	3.8020620 E − 2
-1.5	1.1327886 E − 1	8.9810236 E − 2	7.2556989 E − 2	5.9603004 E − 2	4.9685801 E − 2	4.1958177 E − 2
-1.4	1.2494186 E − 1	9.9070558 E − 2	8.0046200 E − 2	6.5759906 E − 2	5.4821295 E − 2	4.6296929 E − 2
-1.3	1.3777893 E − 1	1.0926593 E − 1	8.8293279 E − 2	7.2540859 E − 2	6.0477950 E − 2	5.1076406 E − 2
-1.2	1.5190283 E − 1	1.2048675 E − 1	9.7371846 E − 2	8.0006696 E − 2	6.6706700 E − 2	5.6339766 E − 2
-1.1	1.6743614 E − 1	1.3283141 E − 1	1.0736208 E − 1	8.8223703 E − 2	7.3563075 E − 2	6.2134082 E − 2
-1.0	1.8451185 E − 1	1.4640680 E − 1	1.1835118 E − 1	9.7264030 E − 2	8.1107547 E − 2	6.8510635 E − 2
-0.9	2.0327410 E − 1	1.6132891 E − 1	1.3043386 E − 1	1.0720610 E − 1	8.9405886 E − 2	7.5525225 E − 2
-0.8	2.2387877 E − 1	1.7772339 E − 1	1.4371287 E − 1	1.1813506 E − 1	9.8529532 E − 2	8.3238490 E − 2
-0.7	2.4649419 E − 1	1.9572618 E − 1	1.5829945 E − 1	1.3014317 E − 1	1.0855597 E − 1	9.1716240 E − 2
0.6	2.7130179 E − 1	2.1548405 E − 1	1.7431389 E − 1	1.4333031 E − 1	1.1956914 E − 1	1.0102979 E − 1
0.5	2.9849674 E − 1	2.3715520 E − 1	1.9188606 E − 1	1.5780439 E − 1	1.3165978 E − 1	1.1125631 E − 1
0.4	3.2828855 E − 1	2.6090981 E − 1	2.1115584 E − 1	1.7368183 E − 1	1.4492590 E − 1	1.2247918 E − 1
-0.3	3.6090164 E − 1	2.8693062 E − 1	2.3227370 E − 1	1.9108797 E − 1	1.5947308 E − 1	1.3478831 E − 1
0.2	3.9657583 E − 1	3.1541340 E − 1	2.5540114 E − 1	2.1015753 E − 1	1.7541496 E − 1	1.4828052 E − 1
0.1	4.3556686 E − 1	3.4656745 E − 1	2.8071115 E − 1	2.3103502 E − 1	1.9287355 E − 1	1.6305987 E − 1
0.0	4.7814669 E − 1	3.8061604 E − 1	3.0838863 E − 1	2.5387515 E − 1	2.1197963 E − 1	1.7923800 E − 1

Numerical values of integral $\phi_3(\xi, b)$

$\xi \downarrow$ / $b \rightarrow$	−1	0	1	2	3	4
0.1	1.5195425	1.7886468	1.5195425	1.1577963	8.7543763 E−1	6.7131498 E
0.2	1.6557293	1.9428590	1.6557293	1.2650603	9.5818645 E−1	7.3557598 E
0.3	1.8023399	2.1080674	1.8023399	1.3810996	1.0479877	8.0545674 E
0.4	1.9599203	2.2847640	1.9599203	1.5064563	1.1453223	8.8136406 E
0.5	2.1290150	2.4734336	2.1290150	1.6416831	1.2506870	9.6372228 E
0.6	2.3101649	2.6745514	2.3101649	Not computed	1.3645938	1.0529726
0.7	2.5039036	2.8885813	2.5039036	1.9439949	1.4875679	1.1495723
0.8	2.7107549	3.1159736	2.7107549	2.1122153	1.6201459	1.2539935
0.9	2.9312300	3.3571641	2.9312300	2.2925707	1.7628738	1.3667224
1.0	3.1658251	3.6125722	3.1658251	2.4856277	1.9163052	1.4882573
1.1	3.4150190	3.8825999	3.4150190	2.6919470	2.0809990	1.6191080
1.2	3.6792715	4.1676312	3.6792715	2.9120814	2.2575168	1.7597935
1.3	3.9590212	4.4680314	3.9590212	3.1465724	2.4464210	1.9108405
1.4	4.2546849	4.7841469	4.2546849	3.3959484	2.6482721	2.0727818
1.5	4.5666559	5.1163043	4.5666559	3.6607219	2.8636266	2.2461546
1.6	4.8953038	5.4648116	4.8953038	3.9413879	3.0930346	2.4314979
1.7	5.2409741	5.8299578	5.2409741	4.2384214	3.3370375	2.6293515
1.8	5.6039879	6.2120131	5.6039879	4.5522764	3.5961659	2.8402536
1.9	5.9846425	6.6112300	5.9846425	4.8833841	3.8709377	3.0647389
2.0	6.3832115	7.0278433	6.3832115	5.2321524	4.1618563	3.3033371
2.1	6.7999458	7.4620711	6.7999458	5.5989645	4.4694087	3.5565705
2.2	7.2350742	7.9141158	7.2350742	5.9841789	4.7940641	3.8249532
2.3	7.6888043	8.3841643	7.6888043	6.3881290	5.1362727	4.1089886
2.4	8.1613241	8.8723892	8.1613241	6.8111231	5.4964645	4.4091686
2.5	8.6528024	9.3789499	8.6528024	7.2534445	5.8750484	4.7259717
2.6	9.1633909	9.9039921	9.1633909	7.7153524	6.2724114	5.0598620
2.7	9.6932245	1.0447651 E+1	9.6932245	8.1970818	6.6889184	5.4112881
2.8	1.0242423 E+1	1.1010049 E+1	1.0242423 E+1	8.6988452	7.1249116	5.7806821
2.9	1.0811094 E+1	1.1591300 E+1	1.0811094 E+1	9.2208324	7.5807104	6.1684585
3.0	1.1399329 E+1	1.2191507 E+1	1.1399329 E+1	9.7632125	8.0566117	6.5750140
3.1	1.2007212 E+1	1.2810765 E+1	1.2007212 E+1	1.0326135 E+1	8.5528897	7.0007265
3.2	1.2634812 E+1	1.3449160 E+1	1.2634812 E+1	1.0909729 E+1	9.0697967	7.4459547
3.3	1.3282194 E+1	1.4106771 E+1	1.3282194 E+1	1.1514107 E+1	9.6075629	7.9110383
3.4	1.3949410 E+1	1.4783671 E+1	1.3949410 E+1	1.2139367 E+1	1.0166398 E+1	8.3962972
3.5	1.4636507 E+1	1.5479923 E+1	1.4636507 E+1	1.2785588 E+1	1.0746490 E+1	8.9020317
3.6	1.5343524 E+1	1.6195590 E+1	1.5343524 E+1	1.3452836 E+1	1.1348010 E+1	9.4285227
3.7	1.6070495 E+1	1.6930724 E+1	1.6070495 E+1	1.4141169 E+1	1.1971106 E+1	9.9760319
3.8	1.6817449 E+1	1.7685374 E+1	1.6817449 E+1	1.4850626 E+1	1.2615913 E+1	1.0544802 E
3.9	1.7584408 E+1	1.8459587 E+1	1.7584408 E+1	1.5581240 E+1	1.3282544 E+1	1.1135056 E
4.0	1.8371393 E+1	1.9253403 E+1	1.8371393 E+1	1.6333033 E+1	1.3971101 E+1	1.1747000 E

Numerical values of integral $\phi_5(\xi, b)$

$\xi \downarrow$ / $b \rightarrow$	−1	0	1	2	3	4
−3.0	2.6471021 E−1	2.9689021 E−1	2.6471021 E−1	2.1388391 E−1	1.6957279 E−1	1.3492454 E
−2.9	2.9239844 E−1	3.2790484 E−1	2.9239844 E−1	2.3628235 E−1	1.8734395 E−1	1.4907121 E
−2.8	3.2296556 E−1	3.6213571 E−1	3.2296556 E−1	2.6101547 E−1	2.0697032 E−1	1.6469618 E
−2.7	3.5670718 E−1	3.9991125 E−1	3.5670718 E−1	2.8832422 E−1	2.2864402 E−1	1.8195282 E
−2.6	3.9394847 E−1	4.4159236 E−1	3.9394847 E−1	3.1847396 E−1	2.5257670 E−1	2.0101023 E
−2.5	4.3504694 E−1	4.8757536 E−1	4.3504694 E−1	3.5175674 E−1	2.7900150 E−1	2.2205471 E
−2.4	4.8039544 E−1	5.3829525 E−1	4.8039544 E−1	3.8849388 E−1	3.0817512 E−1	2.4529150 E
−2.3	5.3042546 E−1	5.9422912 E−1	5.3042546 E−1	4.2903875 E−1	3.4038012 E−1	2.7094665 E

Numerical values of integral $\phi_3(\xi, b)$

$b \rightarrow$	5	6	7	8	9	10
0.1	5.2460386 E−1	4.1779678 E−1	3.3863083 E−1	2.7884323 E−1	2.3287309 E−1	1.9693443 E−1
0.2	5.7524367 E−1	4.5836190 E−1	3.7164762 E−1	3.0611545 E−1	2.5570326 E−1	2.1627692 E−1
0.3	6.3038830 E−1	5.0257856 E−1	4.0766187 E−1	3.3587928 E−1	2.8062925 E−1	2.3740168 E−1
0.4	6.9037675 E−1	5.5072894 E−1	4.4690964 E−1	3.6833368 E−1	3.0782017 E−1	2.6045370 E−1
0.5	7.5556478 E−1	6.0311037 E−1	4.8964036 E−1	4.0368934 E−1	3.3745541 E−1	2.8558699 E−1
0.6	8.2632460 E−1	6.6003530 E−1	5.3611698 E−1	4.4216887 E−1	3.6972485 E−1	3.1296474 E−1
0.7	9.0304452 E−1	7.2183117 E−1	5.8661600 E−1	4.8400695 E−1	4.0482898 E−1	3.4275956 E−1
0.8	9.8612843 E−1	7.8884026 E−1	6.4142742 E−1	5.2945033 E−1	4.4297910 E−1	3.7515359 E−1
0.9	1.0759952	8.6141934 E−1	7.0085466 E−1	5.7875791 E−1	4.8439732 E−1	4.1033865 E−1
1.0	1.1730777	9.3993927 E−1	7.6521438 E−1	6.3220065 E−1	5.2931662 E−1	4.4851627 E−1
1.1	1.2778224	1.0247845	8.3483620 E−1	6.9006147 E−1	5.7798084 E−1	4.8989775 E−1
1.2	1.3906877	1.1163525	9.1006235 E−1	7.5263510 E−1	6.3064461 E−1	5.3470417 E−1
1.3	1.5121435	1.2150529	9.9124728 E−1	8.2022779 E−1	6.8757323 E−1	5.8316631 E−1
1.4	1.6426695	1.3213071	1.0787572	8.9315707 E−1	7.4904247 E−1	6.3552462 E−1
1.5	1.7827540	1.4355469	1.1729694	9.7175137 E−1	8.1533841 E−1	6.9202904 E−1
1.6	1.9328928	1.5582140	1.2742717	1.0563496	8.8675714 E−1	7.5293886 E−1
1.7	2.0935875	1.6897586	1.3830619	1.1473007	9.6360445 E−1	8.1852255 E−1
1.8	2.2653441	1.8306388	1.4997467	1.2449632	1.0461955	8.8905747 E−1
1.9	2.4486717	1.9813189	1.6247412	1.3497043	1.1348545	9.6482965 E−1
2.0	2.6440805	2.1422689	1.7584680	1.4619000	1.2299139	1.0461334
2.1	2.8520810	2.3139628	1.9013562	1.5819336	1.3317147	1.1332712
2.2	3.0731817	2.4968777	2.0538407	1.7101957	1.4406051	1.2265530
2.3	3.3078884	2.6914926	2.2163608	1.8470830	1.5569404	1.3262961
2.4	3.5567022	2.8982870	2.3893601	1.9929979	1.6810825	1.4328245
2.5	3.8201183	3.1177397	2.5732846	2.1483478	1.8133994	1.5464688
2.6	4.0986247	3.3503279	2.7685824	2.3135439	1.9542639	1.6675655
2.7	4.3927011	3.5965259	2.9757024	2.4890009	2.1040542	1.7964564
2.8	4.7028173	3.8568039	3.1950935	2.6751362	2.2631520	1.9334886
2.9	5.0294326	4.1316270	3.4272036	2.8723685	2.4319429	2.0790135
3.0	5.3729943	4.4214542	3.6724788	3.0811180	2.6108156	2.2333865
3.1	5.7339373	4.7267376	3.9313623	3.3018049	2.8001602	2.3969666
3.2	6.1126828	5.0479209	4.2042936	3.5348488	3.0003691	2.5701157
3.3	6.5096381	5.3854393	4.4917079	3.7806684	3.2118356	2.7531981
3.4	6.9251953	5.7397181	4.7940350	4.0396800	3.4349532	2.9465803
3.5	7.3597316	6.1111724	5.1116987	4.3122975	3.6701152	3.1506299
3.6	7.8136080	6.5002060	5.4451160	4.5989315	3.9177141	3.3657156
3.7	8.2871699	6.9072118	5.7946966	4.8999884	4.1781411	3.5922066
3.8	8.7807461	7.3325696	6.1608421	5.2158701	4.4517853	3.8304719
3.9	9.2946488	7.7766470	6.5439456	5.5469735	4.7390333	4.0808800
4.0	9.8291739	8.2397989	6.9443909	5.8936895	5.0402686	4.3437984

Numerical values of integral $\phi_5(\xi, b)$

$\rightarrow$	5	6	7	8	9	10
3.0	1.0851933 E−1	8.8409238 E−2	7.2964662 E−2	6.0964007 E−2	5.1521014 E−2	4.3996963 E−2
2.9	1.1990103 E−1	9.7683780 E−2	8.0620210 E−2	6.7361187 E−2	5.6927794 E−2	4.8614471 E−2
2.8	1.3247286 E−1	1.0792857 E−1	8.9076902 E−2	7.4427990 E−2	6.2900635 E−2	5.3715479 E−2
2.7	1.4635849 E−1	1.1924452 E−1	9.8418120 E−2	8.2234147 E−2	6.9498506 E−2	5.9350371 E−2
2.6	1.6169426 E−1	1.3174291 E−1	1.0873583 E−1	9.0856573 E−2	7.6786458 E−2	6.5574730 E−2
2.5	1.7863048 E−1	1.4554643 E−1	1.2013144 E−1	1.0038009 E−1	8.4836233 E−2	7.2449862 E−2
2.4	1.9733275 E−1	1.6079032 E−1	1.3271673 E−1	1.1089821 E−1	9.3726941 E−2	8.0043373 E−2
2.3	2.1798353 E−1	1.7762359 E−1	1.4661493 E−1	1.2251403 E−1	1.0354579 E−1	8.8429797 E−2

Numerical values of integral $\phi_3(\xi, b)$

$\xi \downarrow$ \ $b \rightarrow$	−1	0	1	2	3	4
−2.2	5.8561059 E−1	6.5589988 E−1	5.8561059 E−1	4.7377981 E−1	3.7592736 E−1	2.9926899 E
−2.1	6.4647030 E−1	7.2388012 E−1	6.4647030 E−1	5.2314379 E−1	4.1515871 E−1	3.3053234 E
−2.0	7.1357398 E−1	7.9879638 E−1	7.1357398 E−1	5.7759927 E−1	4.5844997 E−1	3.6503792 E
−1.9	7.8754521 E−1	8.8133357 E−1	7.8754521 E−1	6.3766040 E−1	5.0621397 E−1	4.0311691 E
−1.8	8.6906636 E−1	9.7223964 E−1	8.6906636 E−1	7.0389096 E−1	5.5890403 E−1	4.4513325 E
−1.7	9.5888345 E−1	1.0723304	9.5888345 E−1	7.7690872 E−1	6.1701758 E−1	4.9148668 E
−1.6	1.0578112	1.1824949	1.0578112	8.5739001 E−1	6.8110007 E−1	5.4261597 E
−1.5	1.1667384	1.3037005	1.1667384	9.4607463 E−1	7.5174917 E−1	5.9900248 E
−1.4	1.2866337	1.4369983	1.2866337	1.0437711	8.2961929 E−1	6.6117386 E
−1.3	1.4185509	1.5835291	1.4185509	1.1513620	9.1542627 E−1	7.2970809 E
−1.2	1.5636358	1.7445288	1.5636358	1.2698097	1.0099525	8.0523778 E
−1.1	1.7231313	1.9213343	1.7231313	1.4001628	1.1140522	8.8845474 E
−1.0	1.8983845	2.1153893	1.8983845	1.5435614	1.2286570	9.8011475 E
−0.9	2.0908525	2.3282499	2.0908525	1.7012443	1.3547818	1.0810427
−0.8	2.3021088	2.5615899	2.3021088	1.8745553	1.4935310	1.1921379
−0.7	2.5338493	2.8172067	2.5338493	2.0649497	1.6461043	1.3143797
−0.6	2.7878985	3.0970258	2.7878985	2.2740010	1.8138037	1.4488332
−0.5	3.0662150	3.4031055	3.0662150	2.5034077	1.9980401	1.5966555
−0.4	3.3708969	3.7376417	3.3708969	2.7549997	2.2003393	1.7591015
−0.3	3.7041870	4.1029707	3.7041870	3.0307448	2.4223494	1.9375305
−0.2	4.0684766	4.5015733	4.0684766	3.3327545	2.6658471	2.1334127
−0.1	4.4663098	4.9360766	4.4663098	3.6632902	2.9327446	2.3483353
0.0	4.9003867	5.4092561	4.9003867	4.0247683	3.2250953	2.5840092
0.1	5.3735650	5.9240370	5.3735650	4.4197649	3.5451008	2.8422754
0.2	5.8888624	6.4834939	5.8888624	4.8510201	3.8951158	3.1251112
0.3	6.4494563	7.0908518	6.4494563	5.3214413	4.2776544	3.4346361
0.4	7.0586846	7.7494834	7.0586846	5.8341059	4.6953941	3.7731180
0.5	7.7200437	8.4629077	7.7200437	6.3922633	5.1511805	4.1429785
0.6	8.4371867	9.2347884	8.4371867	6.9993356	5.6480310	4.5467977
0.7	9.2139208	1.0068930 E+1	9.2139208	7.6589176	6.1891375	4.9873194
0.8	1.0054203 E+1	1.0969274 E+1	1.0054203 E+1	8.3747758	6.7778687	5.4674547
0.9	1.0962137 E+1	1.1939894 E+1	1.0962137 E+1	9.1508466	7.4177713	5.9902859
1.0	1.1941966 E+1	1.2984994 E+1	1.1941966 E+1	9.9912331	8.1125702	6.5590687
1.1	1.2998068 E+1	1.4108899 E+1	1.2998068 E+1	1.0900202 E+1	8.8661688	7.1772355
1.2	1.4134952 E+1	1.5316054 E+1	1.4134952 E+1	1.1882178 E+1	9.6826468	7.8483958
1.3	1.5357246 E+1	1.6611015 E+1	1.5357246 E+1	1.2941738 E+1	1.0566258 E+1	8.5763373
1.4	1.6669696 E+1	1.7998444 E+1	1.6669696 E+1	1.4083609 E+1	1.1521429 E+1	9.3650259
1.5	1.8077156 E+1	1.9483107 E+1	1.8077156 E+1	1.5312653 E+1	1.2552752 E+1	1.0218605
1.6	1.9584581 E+1	2.1069863 E+1	1.9584581 E+1	1.6633870 E+1	1.3664983 E+1	1.1141393
1.7	2.1197020 E+1	2.2763659 E+1	2.1197020 E+1	1.8052381 E+1	1.4863035 E+1	1.2137883
1.8	2.2919612 E+1	2.4569528 E+1	2.2919612 E+1	1.9573427 E+1	1.6151972 E+1	1.3212736
1.9	2.4757574 E+1	2.6492582 E+1	2.4757574 E+1	2.1202356 E+1	1.7537002 E+1	1.4370782
2.0	2.6716199 E+1	2.8538006 E+1	2.6716199 E+1	2.2944617 E+1	1.9023471 E+1	1.5617010
2.1	2.8800848 E+1	3.0711050 E+1	2.8800848 E+1	2.4805751 E+1	2.0616855 E+1	1.6956568
2.2	3.1016944 E+1	3.3017030 E+1	3.1016944 E+1	2.6791384 E+1	2.2322749 E+1	1.8394755
2.3	3.3369967 E+1	3.5461320 E+1	3.3369967 E+1	2.8907217 E+1	2.4146865 E+1	1.9937012
2.4	3.5865448 E+1	3.8049346 E+1	3.5865448 E+1	3.1159020 E+1	2.6095017 E+1	2.1588921
2.5	3.8508965 E+1	4.0786587 E+1	3.8508965 E+1	3.3552623 E+1	2.8173118 E+1	2.3356195
2.6	4.1306138 E+1	4.3678564 E+1	4.1306138 E+1	3.6093909 E+1	3.0387168 E+1	2.5244668
2.7	4.4262625 E+1	4.6730826 E+1	4.4262625 E+1	3.8788809 E−1	3.2743247 E+1	2.7260295
2.8	4.7384118 E+1	4.9949028 E+1	4.7384118 E+1	4.1643294 E+1	3.5247507 E+1	2.9409134
2.9	5.0676341 E+1	5.3338758 E+1	5.0676341 E+1	4.4663371 E+1	3.7906164 E+1	3.1697349
3.0	5.4145043 E+1	5.6905704 E+1	5.4145043 E+1	4.7855071 E+1	4.0725490 E+1	3.4131193
3.1	5.7796000 E+1	6.0655567 E+1	5.7796000 E+1	5.1224463 E+1	4.3711806 E+1	3.6717007

Numerical values of integral $\phi_3(\xi, b)$

$b\rightarrow$	5	6	7	8	9	10
−2.2	2.4078375 E−1	1.9621041 E−1	1.6196178 E−1	1.3534115 E−1	1.1438890 E−1	9.7691290 E−2
−2.1	2.6595465 E−1	2.1673156 E−1	1.7890684 E−1	1.4950473 E−1	1.2636218 E−1	1.0791838 E−1
−2.0	2.9373969 E−1	2.3938608 E−1	1.9761477 E−1	1.6514260 E−1	1.3958227 E−1	1.1921080 E−1
−1.9	3.2440667 E−1	2.6439298 E−1	2.1826683 E−1	1.8240653 E−1	1.5417765 E−1	1.3167838 E−1
−1.8	3.5825010 E−1	2.9199320 E−1	2.4106248 E−1	2.0146355 E−1	1.7028971 E−1	1.4544204 E−1
−1.7	3.9559363 E−1	3.2245164 E−1	2.6622107 E−1	2.2249740 E−1	1.8807403 E−1	1.6063484 E−1
−1.6	4.3679277 E−1	3.5605943 E−1	2.9398378 E−1	2.4571015 E−1	2.0770176 E−1	1.7740317 E−1
−1.5	4.8223781 E−1	3.9313635 E−1	3.2461561 E−1	2.7132390 E−1	2.2936101 E−1	1.9590797 E−1
−1.4	5.3235696 E−1	4.3403343 E−1	3.5840758 E−1	2.9958262 E−1	2.5325852 E−1	2.1632615 E−1
−1.3	5.8761968 E−1	4.7913579 E−1	3.9567913 E−1	3.3075422 E−1	2.7962134 E−1	2.3885202 E−1
−1.2	6.4854032 E−1	5.2886564 E−1	4.3678063 E−1	3.6513264 E−1	3.0869867 E−1	2.6369892 E−1
−1.1	7.1568193 E−1	5.8368554 E−1	4.8209615 E−1	4.0304023 E−1	3.4076386 E−1	2.9110091 E−1
−1.0	7.8966039 E−1	6.4410184 E−1	5.3204637 E−1	4.4483024 E−1	3.7611659 E−1	3.2131466 E−1
−0.9	8.7114872 E−1	7.1066838 E−1	5.8709172 E−1	4.9088947 E−1	4.1508510 E−1	3.5462140 E−1
−0.8	9.6088167 E−1	7.8399039 E−1	6.4773571 E−1	5.4164114 E−1	4.5802871 E−1	3.9132907 E−1
−0.7	1.0596606	8.6472860 E−1	7.1452845 E−1	5.9754790 E−1	5.0534037 E−1	4.3177455 E−1
−0.6	1.1683583	Not computed	7.8807042 E−1	6.5911508 E−1	5.5744949 E−1	4.7632611 E−1
−0.5	1.2879249	1.0514006	8.6901639 E−1	7.2689407 E−1	6.1482485 E−1	5.2538597 E−1
−0.4	1.4193924	1.1589738	9.5807954 E−1	8.0148588 E−1	6.7797773 E−1	5.7939299 E−1
−0.3	1.5638812	1.2772518	1.0560358	8.8354496 E−1	7.4746518 E−1	6.3882554 E−1
−0.2	1.7226056	1.4072424	1.1637285	9.7378305 E−1	8.2389344 E−1	7.0420449 E−1
−0.1	1.8968794	1.5500382	1.2820727	1.0729734	9.0792152 E−1	7.7609640 E−1
0.0	2.0881227	1.7068218	1.4120604	1.1819547	1.0002650	8.5511674 E−1
0.1	2.2978673	1.8788716	1.5547652	1.3016361	1.1016997	9.4193338 E−1
0.2	2.5277630	2.0675671	1.7113475	1.4330010	1.2130660	1.0372701
0.3	2.7795839	2.2743949	1.8830597	1.5771121	Not computed	1.1419101
0.4	3.0552339	2.5009541	2.0712514	1.7351159	1.4693011	1.2567001
0.5	3.3567529	2.7489620	2.2773742	1.9082475	1.6162099	1.3825538
0.6	3.6863222	3.0202598	2.5029877	2.0978356	1.7771386	1.5204560
0.7	4.0462695	3.3168176	2.7497638	2.3053069	1.9533128	1.6714665
0.8	4.4390746	3.6407400	3.0194926	2.5321911	2.1460479	1.8367242
0.9	4.8673736	3.9942709	3.3140867	2.7801257	2.3567538	2.0174508
1.0	5.3339628	4.3797983	3.6355862	3.0508604	2.5869393	2.2149553
1.1	5.8418031	4.7998587	3.9861636	3.3462617	2.8382162	2.4306376
1.2	6.3940226	5.2571412	4.3681280	3.6683168	3.1123031	2.6659923
1.3	6.9939197	5.7544911	4.7839289	4.0191409	3.4110321	2.9226143
1.4	7.6449645	6.2949132	5.2361605	4.4009748	3.7363476	3.2021989
1.5	8.3508009	6.8815746	5.7275650	4.8161950	4.0903151	3.5065496
1.6	9.1152471	7.5178067	6.2610358	5.2673143	4.4751226	3.8375791
1.7	9.9422954	8.2071076	6.8396199	5.7569852	4.8930842	4.1973139
1.8	1.0836112 E+1	8.9531427	7.4665204	6.2880036	5.3466437	4.5878970
1.9	1.1801036 E+1	9.7597458	8.1451004	6.8633105	5.8383773	5.0115913
2.0	1.2841577 E+1	1.0630919 E+1	8.8788795	7.4859953	6.3709963	5.4707825
2.1	1.3962413 E+1	1.1570833 E+1	9.6715404	8.1592968	6.9473498	5.9679818
2.2	1.5168388 E+1	1.2583824 E+1	1.0526926 E+1	8.8866054	7.5704267	6.5058282
2.3	1.6464505 E+1	1.3674397 E+1	1.1449042 E+1	9.6714643	8.2433576	7.0870912
2.4	1.7855927 E+1	1.4847217 E+1	1.2442053 E+1	1.0517570 E+1	8.9694166	7.7146724
2.5	1.9347969 E+1	1.6107112 E+1	1.3510286 E+1	1.1428773 E+1	9.7520221	8.3916075
2.6	2.0946093 E+1	1.7459068 E+1	1.4658225 E+1	1.2409077 E+1	1.0594738 E+1	9.1210680
2.7	2.2655903 E+1	1.8908227 E+1	1.5890514 E+1	1.3462641 E+1	1.1501276 E+1	9.9063620
2.8	2.4483140 E+1	2.0459879 E+1	1.7211952 E+1	1.4593776 E+1	1.2475491 E+1	1.0750936 E+1
2.9	2.6433672 E+1	2.2119462 E+1	1.8627488 E+1	1.5806945 E+1	1.3521388 E+1	1.1658374 E+1
3.0	2.8513494 E+1	2.3892557 E+1	2.0142225 E+1	1.7106762 E+1	1.4643116 E+1	1.2632400 E+1
3.1	3.0728714 E+1	2.5784880 E+1	2.1761410 E+1	1.8497991 E+1	1.5844971 E+1	1.3676879 E+1

Numerical values of integral $\phi_3(\xi, b)$

$\xi \downarrow$ / $b \rightarrow$	-1	0	1	2	3	4
3.2	6.1635010 E+1	6.4594076 E+1	6.1635010 E+1	5.4777617 E+1	4.6871476 E+1	3.9461206 E
3.3	6.5667890 E+1	6.8726984 E+1	6.5667890 E+1	5.8520633 E+1	5.0210896 E+1	4.2370278 E
3.4	6.9900462 E+1	7.3060070 E+1	6.9900462 E+1	6.2459619 E+1	5.3736496 E+1	4.5450771 E
3.5	7.4338618 E+1	7.7599124 E+1	7.4338618 E+1	6.6600691 E+1	5.7454724 E+1	4.8709289 E
3.6	7.8988183 E+1	8.2349965 E+1	7.8988183 E+1	7.0949973 E+1	6.1372051 E+1	5.2152483 E
3.7	8.3855049 E+1	8.7318424 E+1	8.3855049 E+1	7.5513596 E+1	6.5494958 E+1	5.5787047 E
3.8	8.8945107 E+1	9.2510350 E+1	8.8945107 E+1	8.0297687 E+1	6.9829936 E+1	5.9619708 E
3.9	9.4264256 E+1	9.7931605 E+1	9.4264256 E+1	8.5308376 E+1	7.4383480 E+1	6.3657222 E
4.0	9.9818407 E+1	1.0358806 E+2	9.9818407 E+1	9.0551793 E+1	7.9162087 E+1	6.7906370 E

Numerical values of integral $\phi_{9/2}(\xi, b)$

$\xi \downarrow$ / $b \rightarrow$	-1	0	1	2	3	4
−3.0	1.6416403 E−2	4.2619808 E−2	1.6416403 E−2	7.2363997 E−3	3.6256761 E−3	1.9896862
−2.9	1.8114269 E−2	4.6936197 E−2	1.8114269 E−2	7.9895195 E−3	4.0040829 E−3	2.1976686
−2.8	1.9984507 E−2	5.1671520 E−2	1.9984507 E−2	8.8201188 E−3	4.4216525 E−3	2.4272462
−2.7	2.2043926 E−2	5.6862654 E−2	2.2043926 E−2	9.7359743 E−3	4.8823663 E−3	2.6806297
−2.7	2.4310828 E−2	6.2548947 E−2	2.4310828 E−2	1.0745602 E−2	5.3905955 E−3	2.9602494
−2.6	2.6805109 E−2	6.8772223 E−2	2.6805109 E−2	1.1858318 E−2	5.9511351 E−3	3.2687758
−2.4	2.9548362 E−2	7.5576755 E−2	2.9548362 E−2	1.3084305 E−2	6.5692420 E−3	3.6091408
−2.3	3.2563977 E−2	8.3009181 E−2	3.2563977 E−2	1.4434677 E−2	7.2506742 E−3	3.9845622
−2.2	3.5877241 E−2	9.1118376 E−2	3.5877241 E−2	1.5921554 E−2	8.0017332 E−3	4.3985685
−2.1	3.9515424 E−2	9.9955249 E−2	3.9515424 E−2	1.7558128 E−2	8.8293084 E−3	4.8550259
−2.0	4.3507866 E−2	1.0957248 E−1	4.3507866 E−2	1.9358746 E−2	9.7409240 E−3	5.3581673
−1.9	4.7885513 E−2	1.2002415 E−1	4.7885513 E−2	2.1338976 E−2	1.0744788 E−2	5.9126225
−1.8	5.2683046 E−2	1.3136534 E−1	5.2683046 E−2	2.3515685 E−2	1.1849841 E−2	6.5234507
−1.7	5.7935867 E−2	1.4365154 E−1	5.7935867 E−2	2.5906858 E−2	1.3065812 E−2	7.1961736
−1.6	6.3682088 E−2	1.5693807 E−1	6.3682088 E−2	2.8532660 E−2	1.4403269 E−2	7.9368113
−1.5	6.9962021 E−2	1.7127930 E−1	6.9962021 E−2	3.1414040 E−2	1.5873670 E−2	8.7519180
−1.4	8.3146856 E−2	1.8672784 E−1	8.3146856 E−2	3.4573743 E−2	1.7489421 E−2	9.6486195
−1.3	8.4294810 E−2	2.0333360 E−1	8.4294810 E−2	3.8036116 E−2	1.9263924 E−2	1.0634652
−1.2	9.2438444 E−2	2.2114280 E−1	9.2438444 E−2	4.1827149 E−2	2.1211628 E−2	1.1718398
−1.1	1.0129696 E−1	2.4019691 E−1	1.0129696 E−1	4.5974489 E−2	2.3348072 E−2	1.2908927
−1.0	1.1091968 E−1	2.6053155 E−1	1.1091968 E−1	5.0507439 E−2	2.5689930 E−2	1.4216032
−0.9	1.2135689 E−1	2.8217540 E−1	1.2135689 E−1	5.5456937 E−2	2.8255042 E−2	1.5650263
−0.8	1.3265948 E−1	3.0514915 E−1	1.3265948 E−1	6.0855502 E−2	3.1062441 E−2	1.7222963
−0.7	1.4487841 E−1	3.2946448 E−1	1.4487841 E−1	6.6737156 E−2	3.4132368 E−2	1.8946297
−0.6	1.5806416 E−1	3.5512326 E−1	1.5806416 E−1	7.3137313 E−2	3.7486279 E−2	2.0833278
−0.5	1.7226605 E−1	3.8211679 E−1	1.7226605 E−1	8.0092627 E−2	4.1146830 E−2	2.2897785
−0.4	1.8753160 E−1	4.1042535 E−1	1.8753160 E−1	8.7641302 E−2	4.5137856 E−2	2.5154585
−0.3	2.0390571 E−1	4.4001790 E−1	2.0390571 E−1	9.5820388 E−2	4.9484324 E−2	2.7619331
−0.2	2.2142991 E−1	4.7085215 E−1	2.2142991 E−1	1.0467046 E−1	5.4212269 E−2	3.0308566
−0.1	2.4014151 E−1	5.0279276 E−1	2.4014151 E−1	1.1423035 E−1	5.9348710 E−2	3.3239712
0.0	2.6007280 E−1	5.3593987 E−1	2.6007280 E−1	1.2453929 E−1	6.4921544 E−2	3.6431044
0.1	2.8125022 E−1	5.7013873 E−1	2.8125022 E−1	1.3563600 E−1	7.0959412 E−2	3.9901659
0.2	3.0369365 E−1	6.0528869 E−1	3.0369365 E−1	1.4755830 E−1	7.7491543 E−2	4.3671430
0.3	3.2741565 E−1	6.4133876 E−1	3.2741565 E−1	1.6034263 E−1	8.4547581 E−2	4.7760944
0.4	3.5242095 E−1	6.7817521 E−1	3.5242095 E−1	1.7402358 E−1	9.2157379 E−2	5.2191430
0.5	3.7870592 E−1	7.1569917 E−1	3.7870592 E−1	1.8863346 E−1	1.0035078 E−1	5.6984670
0.6	4.0625827 E−1	7.5380927 E−1	4.0625827 E−1	2.0420176 E−1	1.0915736 E−1	6.2159477
0.7	4.3505685 E−1	7.9240319 E−1	4.3505685 E−1	2.2075470 E−1	1.1860621 E−1	6.7748683
0.8	4.6507168 E−1	8.3137927 E−1	4.6507168 E−1	2.3831481 E−1	1.2872562 E−1	7.3764809
0.9	4.9626410 E−1	8.7063789 E−1	4.9626410 E−1	2.5690047 E−1	1.3954281 E−1	8.0234124

Numerical values of integral $\phi_3(\xi, b)$

$b\rightarrow$	5	6	7	8	9	10
3.2	3.3085551 E+1	2.7802280 E+1	2.3490435 E+1	1.9985539 E+1	1.7131392 E+1	1.4795811 E+1
3.3	3.5590329 E+1	2.9950733 E+1	2.5334831 E+1	2.1574462 E+1	1.8506964 E+1	1.5993340 E+1
3.4	3.8249463 E+1	3.2236332 E+1	2.7300264 E+1	2.3269955 E+1	1.9976413 E+1	1.7273744 E+1
3.5	4.1069459 E+1	3.4665290 E+1	2.9392532 E+1	2.5077353 E+1	2.1544604 E+1	1.8641443 E+1
3.6	4.4056905 E+1	3.7243922 E+1	3.1617557 E+1	2.7002126 E+1	2.3216542 E+1	2.0100990 E+1
3.7	4.7218462 E+1	3.9978652 E+1	3.3981386 E+1	2.9049876 E+1	2.4997369 E+1	2.1657074 E+1
3.8	5.0560860 E+1	4.2875996 E+1	3.6490179 E+1	3.1226334 E+1	2.6892357 E+1	2.3314521 E+1
3.9	5.4090889 E+1	4.5942560 E+1	3.9150210 E+1	3.3537355 E+1	2.8906913 E+1	Not computed
4.0	5.7815394 E+1	4.9185036 E+1	4.1967858 E+1	3.5988915 E+1	3.1046568 E+1	2.6953451 E+1

Numerical values of the integral $\phi_{9/2}(\xi, b)$

$b\rightarrow$	5	6	7	8	9	10
-3.0	1.1698231 E-3	7.2634759 E-4	4.7150275 E-4	3.1763105 E-4	2.2080381 E-4	1.5769210 E-4
-2.9	1.2922208 E-3	8.0239220 E-4	5.2088793 E-4	3.5091017 E-4	2.4394351 E-4	1.7422083 E-4
-2.8	1.4273526 E-3	8.8635876 E-4	5.7542250 E-4	3.8766160 E-4	2.6949877 E-4	1.9247566 E-4
-2.7	1.5765276 E-3	9.7906395 E-4	6.3563834 E-4	4.2824449 E-4	2.9771971 E-4	2.1263546 E-4
-2.6	1.7411862 E-3	1.0814068 E-3	7.0212121 E-4	4.7305448 E-4	3.2888189 E-4	2.3489732 E-4
-2.5	1.9229124 E-3	1.1943767 E-3	7.7551591 E-4	5.2252724 E-4	3.6328886 E-4	2.5947839 E-4
-2.4	2.1234474 E-3	1.3190617 E-3	8.5653202 E-4	5.7714233 E-4	4.0127479 E-4	2.8661779 E-4
-2.3	2.3447039 E-3	1.4566582 E-3	9.4595003 E-4	6.3742747 E-4	4.4320754 E-4	3.1657877 E-4
-2.2	2.5887821 E-3	1.6084806 E-3	1.0446280 E-3	7.0396308 E-4	4.8949181 E-4	3.4965104 E-4
-2.1	2.8579863 E-3	1.7759726 E-3	1.1535090 E-3	7.7738722 E-4	5.4057266 E-4	3.8615326 E-4
-2.0	3.1548430 E-3	1.9607186 E-3	1.2736285 E-3	8.5840102 E-4	5.9693929 E-4	4.2643581 E-4
-1.9	3.4821204 E-3	2.1644562 E-3	1.4061232 E-3	9.4777444 E-4	6.5912915 E-4	4.7088375 E-4
-1.8	3.8428484 E-3	2.3890899 E-3	1.5522399 E-3	1.0463525 E-3	7.2773236 E-4	Not computed
-1.7	4.2403409 E-3	2.6367052 E-3	1.7133449 E-3	1.1550619 E-3	8.0339643 E-4	Not computed
-1.6	4.6782186 E-3	2.9095838 E-3	1.8909349 E-3	1.2749185 E-3	8.8683140 E-4	Not computed
-1.5	5.1604327 E-3	3.2102199 E-3	2.0866474 E-3	1.4070345 E-3	9.7881530 E-4	6.9943003 E-4
-1.4	5.6912908 E-3	3.5413368 E-3	2.3022728 E-3	1.5526271 E-3	1.0802000 E-3	7.7193233 E-4
-1.3	6.2754820 E-3	3.9059052 E-3	2.5397664 E-3	1.7130266 E-3	1.1919173 E-3	8.5183559 E-4
-1.2	6.9181044 E-3	4.3071612 E-3	2.8012613 E-3	1.8896862 E-3	Not computed	9.3987180 E-4
-1.1	7.6246925 E-3	4.7486258 E-3	3.0890819 E-3	2.0841908 E-3	1.4505169 E-3	1.0368406 E-3
-1.0	8.4012443 E-3	5.2341243 E-3	3.4057580 E-3	2.2982677 E-3	1.5997240 E-3	1.1436149 E-3
-0.9	9.2542495 E-3	5.7678070 E-3	3.7540392 E-3	2.5337967 E-3	1.7639281 E-3	1.2611461 E-3
-0.8	1.0190716 E-2	6.3541688 E-3	4.1369096 E-3	2.7928208 E-3	1.9445666 E-3	1.3904705 E-3
-0.7	1.1218199 E-2	6.9980699 E-3	4.5576029 E-3	3.0775575 E-3	Not computed	1.5327150 E-3
-0.6	1.2344821 E-2	7.7047561 E-3	5.0196175 E-3	3.3904100 E-3	2.3615286 E-3	1.6891040 E-3
-0.5	1.3579300 E-2	8.4798776 E-3	5.5267319 E-3	3.7339786 E-3	2.6013839 E-3	1.8609651 E-3
-0.4	1.4930969 E-2	9.3295086 E-3	6.0830190 E-3	4.1110723 E-3	2.8647545 E-3	2.0497366 E-3
-0.3	1.6409792 E-2	1.0260164 E-2	6.6928613 E-3	Not computed	3.1537866 E-3	2.2569738 E-3
-0.2	1.8026382 E-2	1.1278816 E-2	7.3609641 E-3	4.9781811 E-3	3.4707026 E-3	2.4842960 E-3
-0.1	1.9792007 E-2	1.2392906 E-2	8.0923688 E-3	Not computed	3.8181688 E-3	2.7336279 E-3
0.0	2.1718596 E-2	1.3610361 E-2	8.8921728 E-3	6.0186342 E-3	4.1987766 E-3	3.0068621 E-3
0.1	2.3818740 E-2	1.4939019 E-2	9.7666915 E-3	6.6135556 E-3	4.6153941 E-3	3.3060899 E-3
0.2	2.6105681 E-2	1.6388924 E-2	1.0721766 E-2	7.2638491 E-3	5.0710901 E-3	3.6336274 E-3
0.3	2.8591985 E-2	1.7969557 E-2	1.1764212 E-2	7.9742677 E-3	Not computed	3.9917246 E-3
0.4	3.1294766 E-2	1.9689601 E-2	1.2900247 E-2	8.7492830 E-3	6.1131532 E-3	4.3830128 E-3
0.5	3.4229156 E-2	2.1560052 E-2	1.4137435 E-2	9.5942170 E-3	6.7066000 E-3	4.8102185 E-3
0.6	3.7408107 E-2	2.3591786 E-2	1.5483405 E-2	1.0514500 E-2	7.3535309 E-3	5.2762395 E-3
0.7	4.0849087 E-2	2.5796138 E-2	1.6946159 E-2	1.1515856 E-2	8.0581062 E-3	5.7841494 E-3
0.8	4.4568675 E-2	2.8184884 E-2	1.8534071 E-2	1.2604305 E-2	8.8247160 E-3	6.3372004 E-3
0.9	4.8583837 E-2	3.0770215 E-2	2.0255874 E-2	1.3786158 E-2	9.6579818 E-3	6.9388261 E-3

Numerical values of integral $\phi_3(\xi, b)$

$\xi \downarrow$ / $b \rightarrow$	−1	0	1	2	3	4
1.0	5.2858710 E−1	9.1008281 E−1	5.2858710 E−1	2.7652557 E−1	1.5108367 E−1	8.7179397 E
1.1	5.6198586 E−1	9.4962224 E−1	5.6198586 E−1	2.9719916 E−1	1.6336231 E−1	9.4623158 E
1.2	5.9639838 E−1	9.8901343 E−1	5.9639838 E−1	3.1892524 E−1	1.7643158 E−1	1.0258753 E
1.3	6.3175625 E−1	1.0284897	6.3175625 E−1	3.4170253 E−1	1.9028117 E−1	1.1109408 E
1.4	6.6798562 E−1	1.0677944	6.6798562 E−1	3.6552437 E−1	2.0493903 E−1	1.2016360
1.5	7.0500808 E−1	1.1069218	7.0500808 E−1	3.9037870 E−1	2.2042031 E−1	1.2981598
1.6	7.4274178 E−1	1.1457764	7.4274178 E−1	4.1624810 E−1	2.3673731 E−1	1.4007003
1.7	7.8110329 E−1	1.1843055	7.8110329 E−1	4.4310992 E−1	2.5389931 E−1	1.5094331
1.8	8.1999255 E−1	1.2224631	8.1999255 E−1	4.7093648 E−1	2.7191247 E−1	1.6245197
1.9	8.5935045 E−1	1.2602099	8.5935045 E−1	4.9964796 E−1	2.9077968 E−1	1.7461061
2.0	8.9907817 E−1	1.2975130	8.9907817 E−1	5.2934958 E−1	3.1050052 E−1	1.8743216
2.1	9.3909179 E−1	1.3343453	9.3909179 E−1	5.5985831 E−1	3.3107124 E−1	2.0092774
2.2	9.7930970 E−1	1.3704651	9.7930970 E−1	5.9117699 E−1	3.5248475 E−1	2.1510658
2.3	1.0196533	1.4056884	1.0196533	6.2325794 E−1	3.7473067 E−1	2.2997592
2.4	1.0600477	1.4417164	1.0600477	Not computed	3.9779546 E−1	2.4554094
2.5	1.1004219	1.4765067	1.1004219	6.8950315 E−1	4.2166247 E−1	2.6180475
2.6	1.1407096	1.5095451	1.1407096	7.2356079 E−1	4.4631212 E−1	2.7876751
2.7	1.1808493	1.5433429	1.1808493	7.5816845 E−1	4.7172210 E−1	2.9642967
2.8	1.2207844	1.5766255	1.2207844	7.9327020 E−1	4.9786751 E−1	3.1478781
2.9	1.2604633	1.6100471	1.2604633	8.2880993 E−1	5.2471889 E−1	3.3383510
3.0	1.2998395	1.6422053	1.2998395	8.6473175 E−1	5.5225365 E−1	3.5356482
3.1	1.3388718	1.6738450	1.3388718	9.0096441 E−1	5.8043386 E−1	3.7396756
3.2	1.3775234	1.7049753	1.3775234	9.3748728 E−1	6.0922897 E−1	3.9503198
3.3	1.4157628	1.7356060	1.4157628	9.7424316 E−1	6.3860483 E−1	4.1674491
3.4	1.4535624	1.7657483	1.4535624	1.0111531	6.6852619 E−1	4.3909148
3.5	1.4908995	1.7954137	1.4908995	1.0481826	6.9895698 E−1	4.6205521
3.6	1.5277551	1.8246142	1.5277551	1.0852843	7.2986049 E−1	4.8561814
3.7	1.5641139	1.8533624	1.5641139	1.1224133	7.6119969 E−1	5.0976093
3.8	1.5999642	1.8816708	1.5999642	1.1595274	7.9293740 E−1	5.3446278
3.9	1.6352974	1.9092546	1.6352974	1.1965791	8.2503563 E−1	5.5970289
4.0	1.6701078	1.9367488	1.6701078	1.2335463	8.5746021 E−1	5.8545783

Numerical values of integral $\phi_3(\xi, b)$

$b \rightarrow$	5	6	7	8	9	10
1.0	5.2911848 E−2	3.3564701 E−2	2.2120650 E−2	1.5068015 E−2	1.0562756 E−2	7.5926430 E−3
1.1	5.7570215 E−2	3.6581257 E−2	2.4137809 E−2	1.6456757 E−2	1.1544120 E−2	8.3024512 E−3
1.2	6.2576591 E−2	3.9833098 E−2	2.6317073 E−2	1.7959540 E−2	1.2607383 E−2	9.0722342 E−3
1.3	Not computed	4.3333693 E−2	2.8668450 E−2	1.9583779 E−2	1.3758075 E−2	9.9061576 E−3
1.4	7.3704156 E−2	4.7096713 E−2	3.1202210 E−2	2.1337142 E−2	1.5001943 E−2	1.0808567 E−2
1.5	7.9860521 E−2	5.1135974 E−2	3.3928852 E−2	2.3227527 E−2	1.6344942 E−2	1.1783984 E−2
1.6	8.6435043 E−2	5.5465376 E−2	3.6859071 E−2	2.5263051 E−2	1.7793226 E−2	1.2837103 E−2
1.7	9.3444625 E−2	6.0098842 E−2	4.0003726 E−2	2.7452025 E−2	1.9353138 E−2	1.3972785 E−2
1.8	1.0090570 E−1	6.5050249 E−2	4.3373799 E−2	2.9802936 E−2	2.1031198 E−2	1.5196050 E−2
1.9	1.0883414 E−1	7.0333363 E−2	4.6980352 E−2	3.2324421 E−2	2.2834088 E−2	1.6512072 E−2
2.0	1.1724511 E−1	7.5961768 E−2	5.0834489 E−2	3.5025243 E−2	2.4768641 E−2	1.7926168 E−2
2.1	1.2615302 E−1	8.1948796 E−2	5.4947309 E−2	3.7914263 E−2	2.6841774 E−2	1.9443759 E−2
2.2	1.3557139 E−1	8.8307458 E−2	5.9329754 E−2	4.1000349 E−2	2.9060657 E−2	2.1070483 E−2
2.3	1.4551276 E−1	9.5050192 E−2	6.3993005 E−2	4.4292604 E−2	3.1432451 E−2	2.2812016 E−2
2.4	1.5598827 E−1	1.0218954 E−1	6.8947840 E−2	4.7799993 E−2	3.3964331 E−2	2.4674098 E−2
2.5	1.6700902 E−1	1.0973710 E−1	7.4204698 E−2	5.1531368 E−2	3.6663644 E−2	2.6662628 E−2
2.6	1.7858407 E−1	1.1770360 E−1	7.9774062 E−2	5.5495695 E−2	3.9537732 E−2	2.8783556 E−2
2.7	1.9072068 E−1	1.2609962 E−1	8.5666040 E−2	5.9701796 E−2	4.2593951 E−2	3.1042887 E−2
2.8	2.0342577 E−1	1.3493486 E−1	9.1890412 E−2	6.4158380 E−2	4.5839650 E−2	3.3446672 E−2
2.9	2.1670454 E−1	1.4421829 E−1	9.8456590 E−2	6.8874005 E−2	4.9282146 E−2	3.6000993 E−2
3.0	2.3056085 E−1	1.5395805 E−1	1.0537358 E−1	7.3857050 E−2	5.2928705 E−2	3.8711944 E−2
3.1	2.4499712 E−1	1.6416148 E−1	1.1264993 E−1	7.9115683 E−2	5.6786517 E−2	4.1585616 E−2
3.2	2.6001438 E−1	1.7483502 E−1	1.2029373 E−1	8.4657834 E−2	6.0842441 E−2	4.4628083 E−2
3.3	2.7561224 E−1	1.8598422 E−1	1.2831251 E−1	9.0466011 E−2	6.5134524 E−2	4.7845381 E−2
3.4	2.9178891 E−1	1.9761368 E−1	1.3668197 E−1	9.6623022 E−2	6.9697781 E−2	5.1243509 E−2
3.5	3.0854123 E−1	2.0972709 E−1	1.4550246 E−1	1.0306047 E−1	7.4470239 E−2	5.4828385 E−2
3.6	3.2586467 E−1	2.2232710 E−1	1.5468590 E−1	1.0981022 E−1	7.9488020 E−2	5.8605857 E−2
3.7	3.4375328 E−1	2.3541561 E−1	1.6426881 E−1	1.1687855 E−1	8.4757393 E−2	6.2581658 E−2
3.8	3.6220031 E−1	2.4899326 E−1	1.7425573 E−1	1.2427142 E−1	9.0284434 E−2	6.6761426 E−2
3.9	3.8119709 E−1	2.6305991 E−1	1.8465055 E−1	1.3199435 E−1	9.6074968 E−2	7.1150664 E−2
4.0	4.0073426 E−1	2.7761444 E−1	1.9545652 E−1	1.4005243 E−1	1.0213457 E−1	7.5754736 E−2

REFERENCES

1. E.S. Rittner, 'On the theory of the Peltier Heat Pump', *Journal of Applied Physics*, **30** (1959), 702–7.
2. R.P. Chasmar and R.J. Stratton, 'The thermoelectric figure of merit and its relation to thermoelectric generators', *Journal of Electronics and Control*, **7** (1959), 52–72.
3. H. Littman and B. Davidson, 'Theoretical bound on the thermoelectric figure of merit from irreversible thermodynamics', *Journal of Applied Physics*, **32** (1961), 217–19.
4. R. Simon, 'Thermoelectric figure of merit of two band semiconductors', *Journal of Applied Physics*, **33** (1962), 1830–41.
5. E.S. Rittner and G.F. Neumark, 'Theoretical bound on the thermoelectric figure of merit of two-band semiconductors', *Journal of Applied Physics*, **34**, (1963), 2071–7.
6. R.W. Ure, Jr, 'Practical limits to the thermoelectric figure of merit, II' *Energy Conversion*, **12** (1972), 45–52.
7. V.I. Fistul, *Heavily Doped Semiconductors*, Plenum, New York, 1969.
8. H.J. Goldsmid, A.R. Sheard, and D.A. Wright, 'The performance of bismuth telluride thermojunctions'. *British Journal of Applied Physics*, **9** (1958), 365–70.

5

Thermoelectric Materials

5.1 FACTORS INFLUENCING CHOICE OF THERMOELECTRIC MATERIALS

5.1.1 Introduction

ALL three parameters occurring in the figure of merit are functions of the carrier concentration, usually expressed in terms of reduced Fermi energy ξ. In silicon–germanium alloys at room temperature, $\xi=0$ corresponds to $\sim 10^{25}$ m^{-3}, and $\xi=4$ to about 10^{26} m^{-3}. The expected dependency of σ, α and λ are shown schematically in Figure 5.1. The figure of merit reaches a maximum at carrier concentration values in the range 10^{25} to 10^{26} m^{-3} i.e., in heavily doped semiconductors.

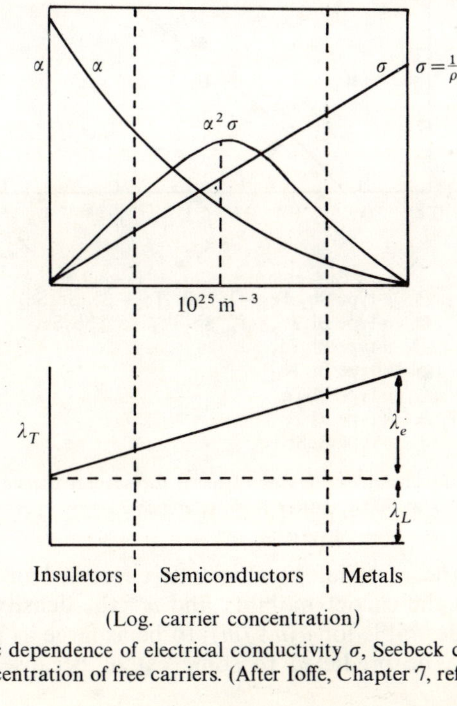

Figure 5.1 Schematic dependence of electrical conductivity σ, Seebeck coefficient α, and thermal conductivity λ on concentration of free carriers. (After Ioffe, Chapter 7, reference 19.)

5.1.2 The Material Parameter

Although some interest has been retained in metals and metallic binary continuous solid solutions as thermoelectric materials, the absence of a physical theory of the thermoelectric properties of metals has resulted in slow progress being made in this direction[1]. Solid state theory, on the other hand, assists in the search amongst semiconductors for potentially good thermoelectric materials inasmuch as it allows the representation of the figure of merit in terms of fundamental properties through a so-called material parameter β. The thermal conductivity λ is the sum of a contribution arising from lattice waves (phonons) λ_L and a contribution λ_e from the carriers present. In a metal λ_L is simply related to λ_e by the Wiedemann–Franz–Lorenz law, and a similar but rather more complicated relationship exists for heavily doped semiconductors. Once the Seebeck coefficient has been optimized by suitable doping, the maximum thermoelectric figure of merit can be expressed in terms of the λ_e/λ_L ratio as indicated in Figure 5.2.[2]

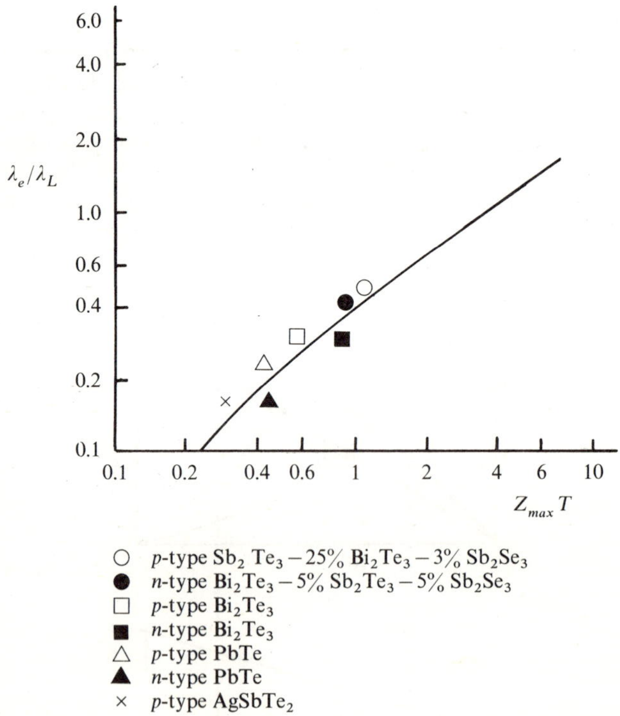

Figure 5.2 Maximum dimensionless figures of merit for several thermoelectric materials as a function of the ratio λ_e/λ_L at 300 K. (After Rosi, reference 2.)

Conventionally, the material parameter is expressed in the form $\beta = \mu/\lambda_L (m^*/m)^{\frac{3}{2}}$ where μ is the carrier mobility and m^* the density of states effective mass. Clearly, it is desirable for $\mu(m^*/m)^{\frac{3}{2}}$ to be as large as possible and for λ_L to be small. λ_L can be predicted to some extent because the thermal conductivity depends on the type of bond and on the atomic weight. Generally, thermal conductivity increases with hardness and decreases with increase in atomic weight. Consequently materials exhibiting covalent bonding will have high values of thermal conductivity, whereas ionic crystals will be poor conductors of heat. The thermal conductivities of a number of elements and compounds are plotted in Figure 5.3 as a function of mean atomic weight and

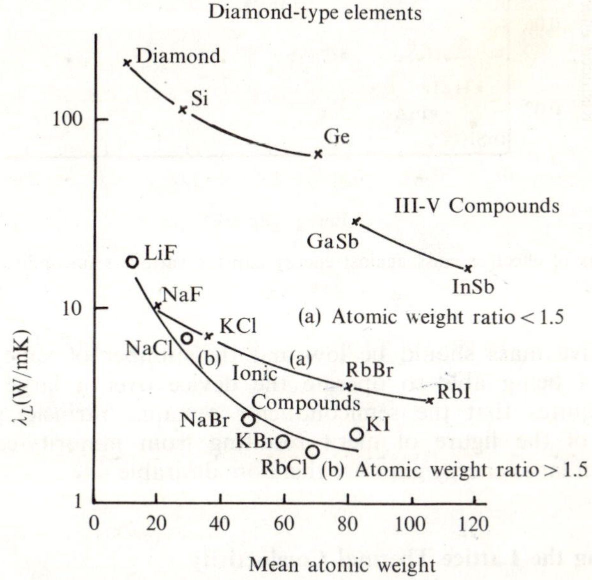

Figure 5.3 Lattice thermal conductivity as a function of mean atomic weight and bonding. (After Ioffe and Ioffe, reference 3.)

bonding.[3] Unfortunately the carrier mobility is also dependent on the type of bonding, being highest in covalent and lowest in ionic semiconductors. Nevertheless, desirable materials are those that possess the highest mean atomic weight. The behaviour of $\mu(m^*/m)^{\frac{3}{2}}$ is more difficult to predict. The mobility is related to the effective mass by $\mu = \tau e/m_c$ where m_c is the conductivity (inertial) effective mass and τ the relaxation time. The effective mass depends upon the energy band structure and the energy gap (in fact the energy gap itself is included in the band structure). Plots of effective mass against energy gap for various semiconductors are shown in Figure 5.4.[4] The lowest value of effective mass is that of indium antimonide: $0.014\,m$ where m is the free electron mass. From the figure we see that a mass this small can be expected only if the energy gap is of the order of 0.01 eV. However, with a gap this small there would be appreciable hole density, particularly at high temperature, and a two-band semiconductor model needs to be employed. In the simplest semiconductor model m_c and m^* are equal, but normally thermoelectric semiconductor models are complex with energy surfaces having a number of valleys, N_v, at the same energy; the effective mass is then given by $m^* N_v^{-\frac{2}{3}}$.

As a rule of thumb, the material parameter is proportional to N_v/m_c, i.e., the

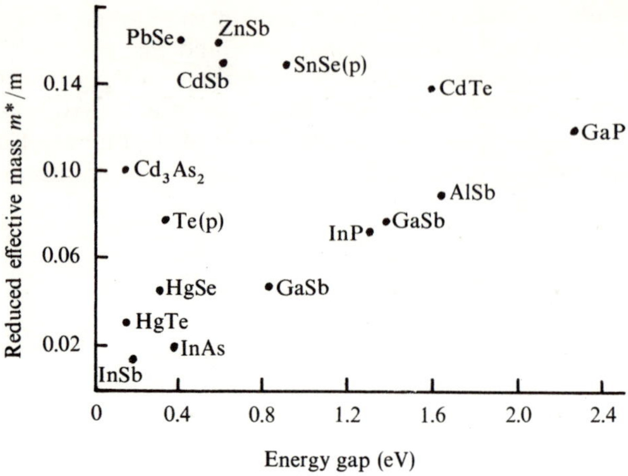

Figure 5.4 Plots of effective mass against energy gap for various semiconductors. (After Ure, reference 4.)

inertial effective mass should be low and the number of valleys high. The desirability of being able to operate the device over a large temperature difference requires that the semiconductor remains intrinsic, so inhibiting degradation of the figure of merit resulting from minority-carrier effects. Materials with large energy gaps are therefore desirable.

5.1.3 Reducing the Lattice Thermal Conductivity

Substantial improvement to the thermoelectric figure of merit accompanies the use of solid solutions[5] when the thermoelectric material is alloyed with an isomorphous element or compound. In these materials the long-range periodicity of the lattice is maintained, but the short-range distortion—resulting from different size atoms in adjacent lattice sites—is effective in scattering the relatively short wavelength of phonons but is ineffective in scattering the orders of magnitude longer wavelengths associated with charge carriers. The effect of alloying on the thermal resistivity of some semiconductors is shown in Figure 5.5.[5] In addition to reducing the thermal conductivity, alloying facilitates the tailoring of the energy gap to meet the envisaged practical requirements of the device.

Silicon–germanium is an established high-temperature thermoelectric material employed in both 'single crystal' and 'compacted' form. The lattice thermal conductivity of the alloy may be further reduced in fine grained materials, an effect attributed to phonon scattering at grain boundaries.[6,7,8,9] Figure 5.6 shows the lattice thermal conductivity and energy gap for Si–Ge alloy along with those of other important semiconductors.[10] The beneficial effect of alloying, doping, and hot pressing in reducing the lattice thermal conductivity of silicon–germanium is apparent. Although silicon–germaniun alloy is remarkable in possessing a large energy gap and low thermal conductivity, its realization of an outstanding figure of merit is thwarted by its large carrier effective mass and accompanying low carrier mobility.

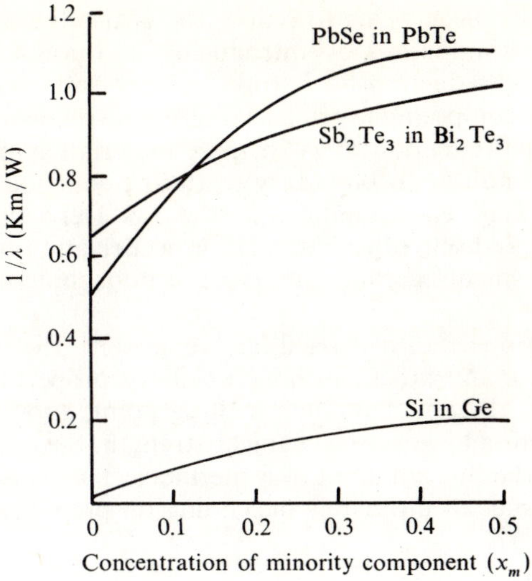

Figure 5.5 The effect of alloying on thermal resistivity of some semiconductors. (After Ioffe et al., reference 5.)

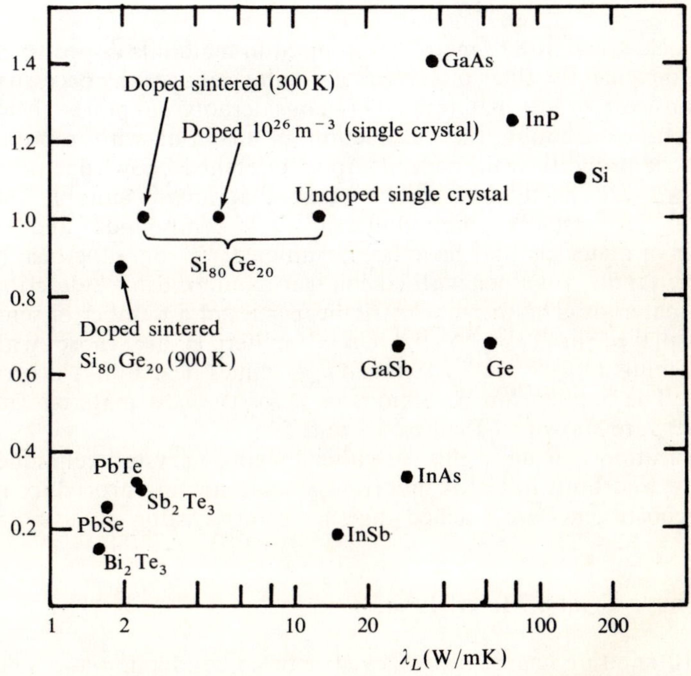

Figure 5.6 Lattice thermal conductivity and energy gap of silicon–germanium alloy system along with those of other important semiconductors. Values correspond to room temperature unless stated otherwise. (After Bhandari and Rowe, reference 10.)

Attempts have also been made to reduce the lattice thermal conductivity of thermoelectric semiconductors by introducing additional disorder into the structure. Substantial reductions in λ_L (up to 50 per cent) have been reported[11] when Group III–V compounds such as gallium phosphide are added to Si–Ge alloys. However, the reduction in λ_L in these materials is probably enhanced by a significant phonon-grain boundary scattering contribution. Nevertheless, initial results are very encouraging and the US Department of Energy is funding two major and one other research project relating to the development and assessment of silicon–germanium–gallium phosphide and similar materials.

Although the properties discussed above govern the performance of a material, there are many other desirable ancillary properties that need to be taken into account; these include high melting point, good chemical stability, low vapour pressure and good mechanical strength. Several of these conflict, and realization of the full potential of a thermoelectric material is a matter of compromise, with success ultimately depending on preparation and doping to optimum level.

5.2 PROPERTIES AND PREPARATION OF ESTABLISHED MATERIALS

5.2.1 Introduction

Thermoelectric figure of merit is optimized in materials doped to about 0.1 per cent. Conveniently, their high level of doping removes the necessity of employing transistor grade materials. This considerably simplifies their method of preparation, although the production of material with reproducible thermoelectric properties still depends upon a detailed knowledge of its solid state chemistry and particularly of its phase diagram.[12] Suitable thermoelectric materials are usually compounds or alloy compounds, and of the great number of materials that have been examined[13,14] only the bismuth and lead telluride family, together with silicon–germanium, can be described as established materials. The thermoelectric properties of a number of semiconductors are displayed in Table 5.1.[15] Useful materials are those with Z values approaching $3 \times 10^{-3}\,K^{-1}$, giving ZT_{max} values near unity; others are clearly useless. The temperature variations of Z for the best materials, including the selenides, are shown in Figures 5.7 and 5.8.

Preparation is usually effected either by single crystal techniques or by hot pressing, and both methods involve a similar starting procedure in which the initial constituents are reached together before casting.

5.2.2 Single Crystal Materials

Bismuth telluride has a multivalley structured conduction and valence band[16] with an indirect bandgap of 0.15 eV;[17] it is essentially a low temperature thermoelectric material with a maximum operating temperature of around 450 K. It is the only one of the established materials that exhibits anisotropic

Table 5.1 *Thermoelectric properties of a number of semiconductors* (after Wright, reference 15)

1	2 E_g (eV)	3 Melting point (K)	4 Type of semiconductor	5 $\alpha^2\sigma$ (W/K^{2}m)	6 λ_e (W/Km)	7 λ_L (W/Km)	8 Z_{max} (K^{-1})	9 Temperature for columns 5–8 (K)	10 Maximum operating temperature (K)
Bi_2Te_3	0.15	848	n or p	4×10^{-3}	0.4	1.6	2×10^{-3}	300	450
$BiSb_4Te_{7.5}$	–	–	p	4.6×10^{-3}	0.4	1.0	3.3×10^{-5}	300	450
$BiTe_2Se$	0.3	–	n	3.6×10^{-3}	0.3	1.3	2.3×10^{-5}	300	600
PbTe	0.3	1177	n or p	2.6×10^{-3}	0.3	2.0	1.2×10^{-3}	300	900
GeTe(+Bi)	–	998	p	3.2×10^{-3}	0.6	1.6	1.6×10^{-3}	800	900
ZnSb	0.6	819	p	2×10^{-3}	–	17.0	1.2×10^{-3}	500	600
$AgSbTe_2$	0.6	849	p	1.4×10^{-3}	0.3	0.6	1.8×10^{-3}	700	900
InAs(+P)	0.45	1213	n	4×10^{-3}	0.5	6.5	6×10^{-4}	900	1100
$CuTe_3S$	–	1203	–	1.8×10^{-3}	–	1.2	1.5×10^{-3}	1100	–
Ge	0.65	1231	n or p	4×10^{-3}	0.6	63	6.3×10^{-5}	300	–
Si	1.15	1693	n or p	2×10^{-3}	0.3	113	1.5×10^{-5}	300	–
SiGe	0.8	–	n	3×10^{-3}	0.3	3.3	9×10^{-4}	900	1200
			p	2×10^{-3}			6×10^{-4}		
InSb	0.17	800	n	8×10^{-3}	1.0	16.0	4.7×10^{-4}	300	–

The values of $\alpha^2\sigma$ in column 5 are the maximum obtainable by optimum doping to give $\alpha = 200$–$250\,\mu VK^{-1}$. The correct value of α in this range is higher, the higher is λ/λ_L, and is near 250 for AgSbTe and 200 for InAs.
The values for λ_e in column 6 apply to the case of optimum doping.

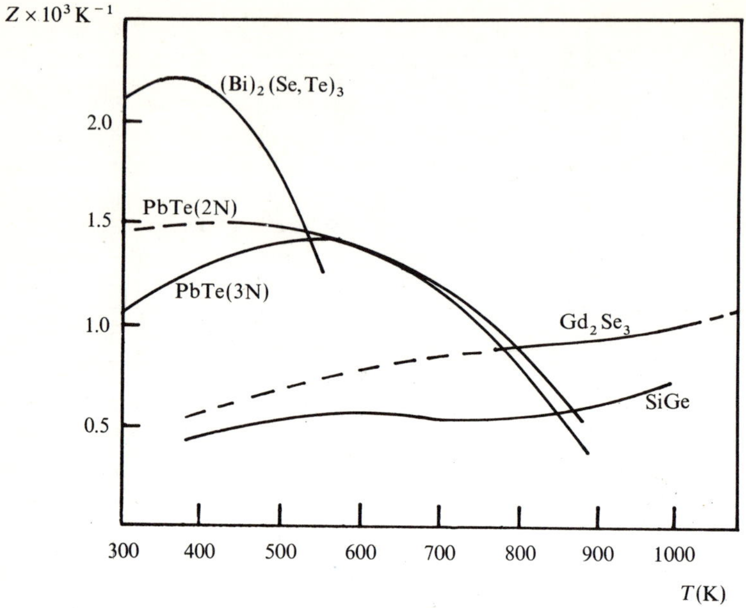

Figure 5.7 Temperature variation of the thermoelectric figure of merit for *n*-type alloys. (After Bass and Elsner, reference 37.)

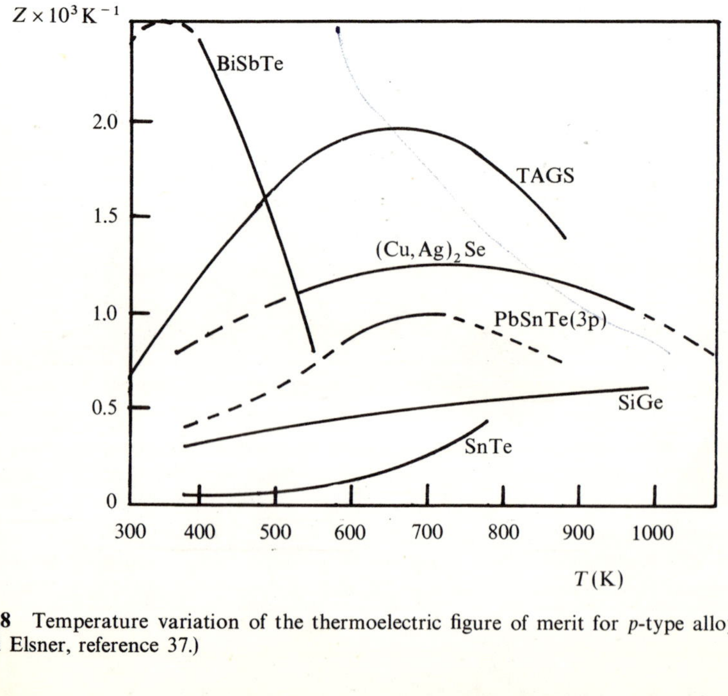

Figure 5.8 Temperature variation of the thermoelectric figure of merit for *p*-type alloys. (After Bass and Elsner, reference 37.)

properties with maximum Z values for both n- and p-type material obtained with current flow parallel to the cleavage planes.[18]

Single-crystal material is prepared by zone levelling or by the loaded zone process.[19] The loaded zone process offers a bonus in that the cleavage planes of the individual crystals, and hence the directions in which maximum Z values are obtained, are formed parallel to the zone-levelled ingot. Bismuth telluride can be doped n-type with halogens, silver or copper (although in practice silver and copper are not used because of their fast diffusion) and p-type with cadmium, tin or lead. Its lattice thermal conductivity can be reduced by forming isomorphous alloys with bismuth selenide or antimony telluride, which are n- and p-type respectively. Minimum λ_L values for $(Bi–Sb)_2Te_3$ occur near 70 per cent Sb, and both $BiSbTe_3$ and $Bi_{0.5}Sb_{1.5}Te_3$ are used as p-type materials.

Lead telluride and its alloys possess a cubic structure and exhibit isotropic electric properties. Lead telluride's energy gap is 0.3 eV and its maximum temperature of operation is considerably higher than that of the bismuth telluride family. It can be made n- or p-type as a result of departure from stoichiometry, but to optimize the Seebeck coefficient doping is necessary. As with bismuth telluride, doping is achieved by the addition of halogens to produce n-type material or of alkali metals to produce p-type materials. Single crystal material is produced either by the Bridgeman or Stockburger methods.[19] Materials developed by the 3 M company are identified by designations such as TEGS-2N. This indicates a specific semiconductor composed essentially of a compound of lead and tellurium with small additions of an electrically active impurity (0.03 mol per cent PbI_2) in order to provide the desired concentration of electrical carriers (n-type), while TEGS-3N is doped n-type with 0.055 mol per cent PbI_2. On the other hand -2P signifies that the lead telluride is doped with Na to give p-type material, while -3P signifies a lead–tin telluride combination doped with sodium and manganese.

The lattice thermal conductivity can be reduced by the addition of SnTe, and this reduction more than offsets any increase in electrical resistivity that has resulted from a fall in carrier mobility. Unfortunately the improvement is offset by a reduction in energy gap, and the material is optimized at a composition of $Pb_{0.75}Sn_{0.25}Te$.[20] More complex materials such as TAG-85 (silver–tin–tellurium) and TAGS (Tellurium–antimony–germanium–silver) have also been developed.[21]

Although individually silicon and germanium are of no use in thermoelectric application because of their high value of λ_L, an alloy of the two displays a marked reduction in λ_L and exhibits a broad maximum in $1/\lambda$ near the middle of the alloy system[22,23,24] (Figure 5.9). The melting point and band gap of the alloy increases with silicon content and consequently alloys of composition $Si_{85}Ge_{15}$ and $Si_{70}Ge_{30}$ have been most extensively studied. The maximum of the Seebeck coefficient occurs at the composition where the conduction bands of germanium and silicon become equi-energetic (Figure 5.10). As can be seen from Figures 5.7 and 5.8, these materials are among the best for high temperature operation. Single-crystal material is prepared by zone-levelling.[25] Inhomogeneity in alloy composition due to constitutional supercooling[26] is inhibited by careful choice of growth rates, and silicon–germanium alloy with less than 1 per cent variation in alloy composition can be prepared. Si–Ge alloys suitable for thermoelectric applications can also be produced by the Wacker process.[27]

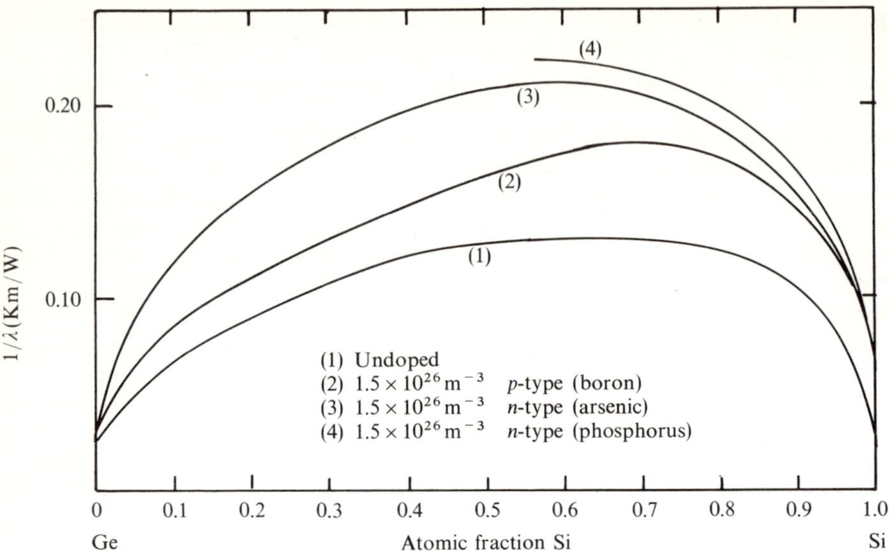

Figure 5.9 Thermal resistivity of undoped and doped *p*- and *n*-type silicon–germanium alloys as a function of alloy composition at 300 K. (After Rosi, reference 2.)

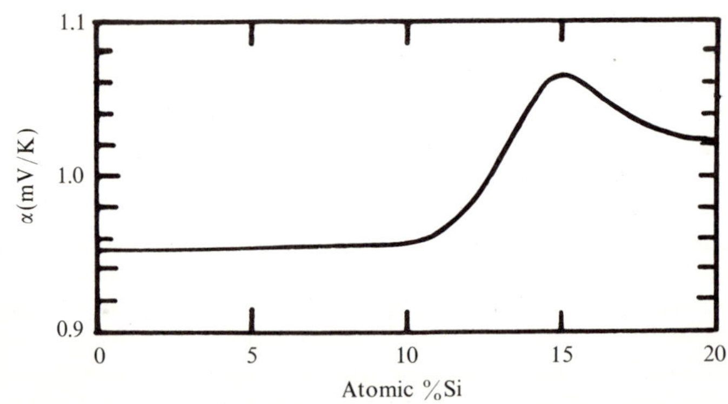

Figure 5.10 Variation of Seebeck coefficient with composition in lightly doped silicon–germanium alloys ($n \sim 1.3 \times 10^{21}\ \mathrm{m}^{-3}$) at room temperature. (After Rosi, reference 2.)

In order to approach optimum material performance the silicon–germanium alloys are doped to the solid solubility limits of the dopants within the alloys. The dopants possess retrograde characteristics that are strongly temperature dependent.[28] Consequently, the amount of dopant in solution at a particular temperature can exceed the solid solubility limit at that temperature, and dopant precipitates out of solution as a function of time.[29] This can significantly affect the transport properties. An overview of the effect of phosphorus precipitation on the room temperature Seebeck coefficient during isothermal heat treatment of $Si_{63.5}$ $Ge_{36.5}$ alloy is shown in Figure 5.11.[30] The precipitation of phosphorus is very fast over the temperature range 700 to 800 K,

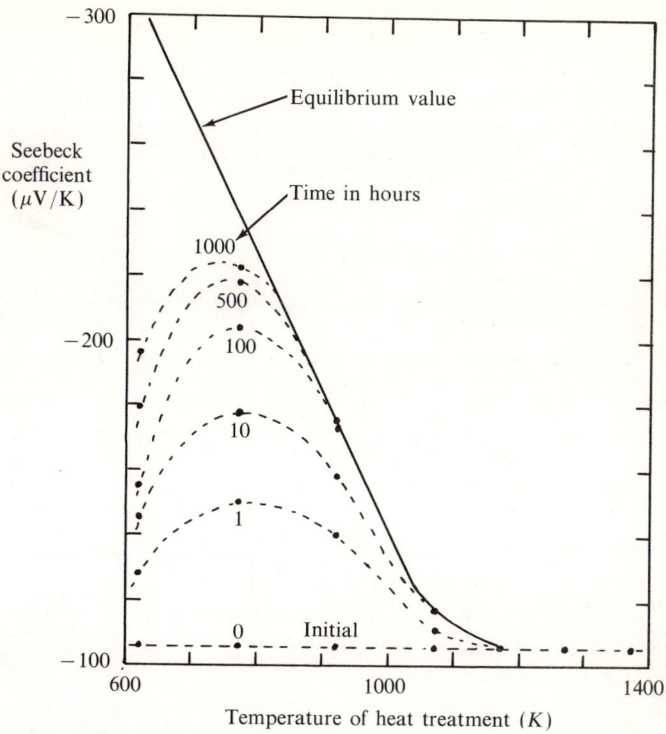

Figure 5.11 Effect of phosphorus precipitation on the Seebeck coefficient of $Si_{63.5}Ge_{36.5}$ alloy. (After Savvides and Rowe, reference 30.)

with the carrier concentration reducing by ~50 per cent after only one hour. However, the process is reversible and the dopant can be put back in solution by suitable annealing. The precipitation of boron is considerably slower than that of phosphorus and is fastest in the temperature range 1000 to 1200 K.[31] Considerable success has been enjoyed in applying the Lifshitz–Slyozov model to the precipitation process, and the long-term behaviour of the thermoelectric transport properties of silicon–germanium alloys can be predicted (Figure 5.12) from the results of relatively short-term heat treatment.[32]

The selenides employed in thermoelectric application have been given the designation TPM-217. Typical members of this class include *p*-type materials composed of copper, silver and selenium in proportions similar to $Cu_{1.97}Ag_{0.03}Se_{1.0081}$, and *n*-type materials composed of gadolinium and selenium with compositions similar to $GdSe_{1.49}$. Good thermoelectric properties are derived from a disordered lattice structure and a non-stoichiometric doping mechanism. The disordered lattice structure is advantageous in that it results in a short phonon mean free path and a consequent low lattice thermal conductivity. It also results in these compounds existing over a wide range of compositions, and facilitates the doping of the material by change of stoichiometry rather than through the conventional addition of a third element. The following relationships hold between extrinsic carrier concentration and

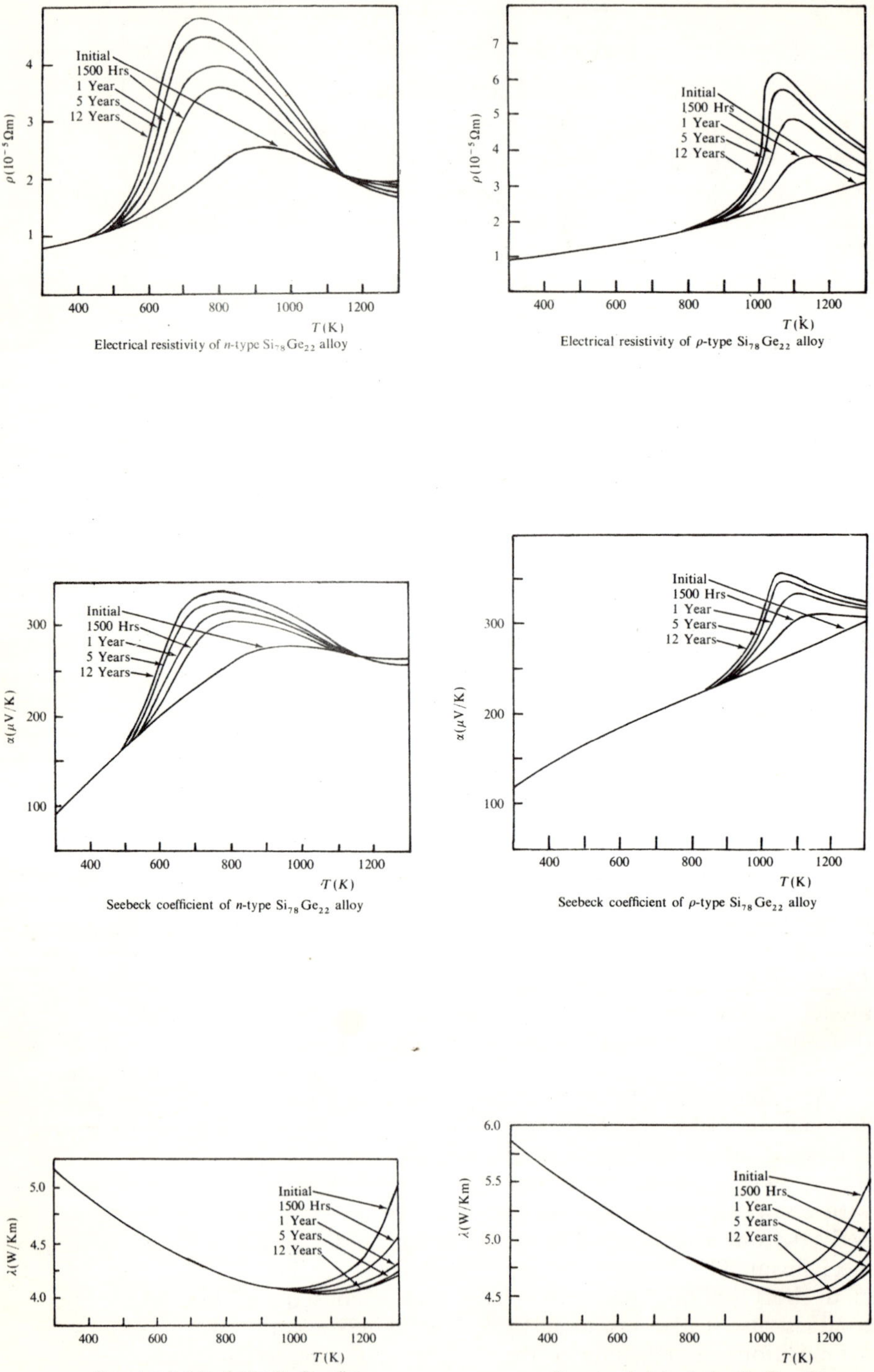

Figure 5.12 Long-term behaviour of silicon–germanium alloys. (After Raag, reference 32.)

composition:

(1) $n_h = 4 \times 10^{28}\, y$ holes/m^3 for $Cu_{1.97}\,Ag_{0.03}\,Se_{1+y}$.

(2) $n_e = 3.33 \times 10^{28}\, y$ electrons/m^3 for $GdSe_{1.5-y}$.

For both *p*- and *n*-type materials these linear relations have been found to hold over the range of thermoelectric interest. The convention has been adopted of expressing the carrier concentration n_h for the *p*-type material in terms of atomic per cent excess selenium '*n*' using equation 1 and the additional relation: atomic per cent selenium $= 22.2y$.

The dependence of the Seebeck coefficient and electrical resistivity of *p*-type $Cu_{1.97}\,Ag_{0.03}\,Se_{1+y}$ is shown in Figures 5.13 and 5.14 as a function of

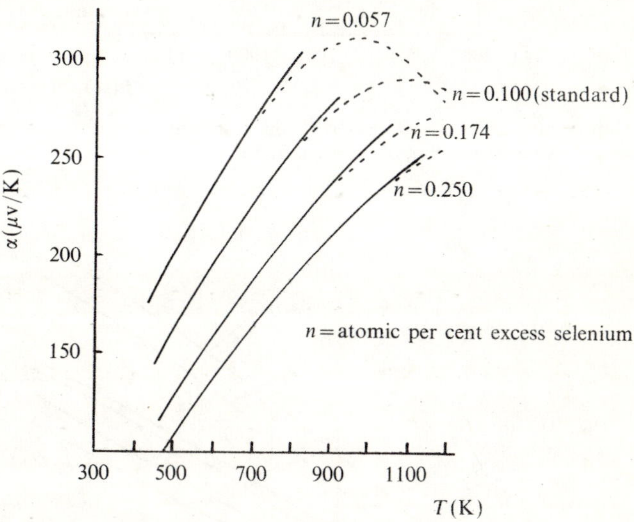

Figure 5.13 Dependence of Seebeck coefficient on composition and temperature for *p*-type $Cu_{1.97}Ag_{0.03}Se_{1+y}$. (After Mitchell et al., reference 33.)

temperature.[33] Although it is the low thermal conductivity of the selenides which is largely responsible for their superior thermoelectric performance, information on this parameter is scant.[34] An interesting and exciting property of *p*-type TPM-217 results from the high mobility of a major constituent, copper. The relatively short response time of copper to electric and thermal fields results in a steady state being quickly reached in which the concentration of dopant increases significantly from the cold to the hot end. This effectively serves to optimize the average leg efficiency.[35] The preparation of the gadolinium metal material with constant properties presents a major problem in the use of this material.[36] In addition, although the selenides possess the highest figure of merit in the 800 to 1200 K temperature range[37] of any other known material, considerable difficulties are encountered if operated below 550 K[38] for a variety of reasons. Consequently, attempts are being made to segment *p*-type material with BiSbTe and the *n*-type with 2N-PbTe.[37] A comparison of the potential thermoelectric efficiency of paired thermoelectric materials is given in Figure 5.15.[35]

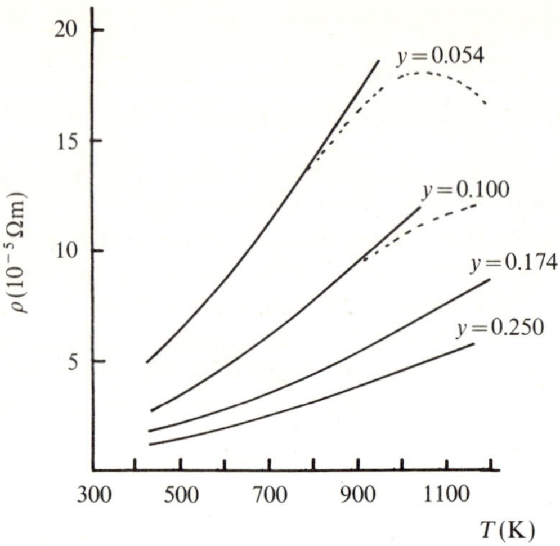

Figure 5.14 Dependence of electrical resistivity on composition and temperature for *p*-type $Cu_{1.97}Ag_{0.03}Se_{1+y}$. (After Mitchell et al., reference 33.)

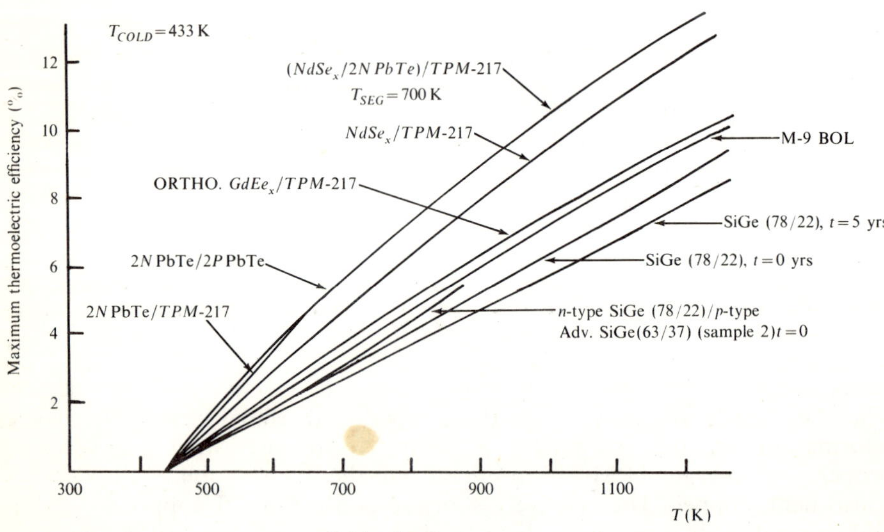

Figure 5.15 Potential thermoelectric efficiency of paired thermoelectric materials. (After Hinderman et al., reference 35.)

Characteristics of the thermoelectric properties of TPM-217 are incomplete.[39,40] Nevertheless, this material was chosen for use in the low cost high performance generator (LCHPG) technology programme initiated by the US Air Force in 1973. These materials were in competition with silicon–germanium alloys for use in the Galileo and International Solar Polar missions. Serious difficulties were encountered in producing stable selenide thermoelectrics and it has recently been confirmed that generators based upon silicon–germanium alloys will be used on these missions.[41,42]

5.2.3 Hot Pressed Materials

The hot-pressing method of preparation is attractive in that it facilitates device fabrication and can readily be adapted to large scale production of material (see Section 6.3). Cast material of the required composition and dopant concentration is ball milled or ground and then sieved, and the required sieved fraction is taken to the next process.[43] The method of compaction varies from one material to another. For bismuth telluride and its alloys, the sieved fraction with a size range 5 to 200 μm is initially cold pressed into discs of the required size, followed by sintering for 1 hour at about 670 K in an atmosphere of hydrogen.[44,45] PbTe and GeTe are prepared by cold pressing; silicon–germanium is hot pressed at about 1700 K.[43,44,46,47,48,49]

Attempts are being made to develop materials suitable for operation at temperatures of 1500 to 1800 K.[50] Based upon hot pressed alloys of boron–silicon–carbon (*p*-type) and of the rare earth sulphides, particularly the sequisulphides of lanthanum, praseodymium and cerium (*n*-type), the objective is to produce a material capable of achieving 15 per cent generating efficiency in vacuum conditions.

As previously indicated, materials prepared by hot pressing methods hold out a possibility of effecting a reduction in thermal conductivity by grain boundary scattering. This phenomenon is most pronounced in SiGe alloys where the constituent elements have large differences in atomic masses. In these alloys, short wavelength phonons are scattered by alloy disorder, which results in the heat being conducted by phonons of long wavelength which are effectively scattered at grain boundaries. Significant advances have been made in this area and good agreement reached between the predicted reduction in the lattice thermal conductivity with grain size and experimental data as shown in Figure 5.16.[51] Measurements of the temperature variation of the lattice thermal conductivity of $Si_{63.5}$ $Ge_{36.5}$ alloy for different grain sizes is shown in Figure 5.17.[52] The lattice thermal conductivity of the 'single crystal material' is equal to 4.31 W/Km. This decreases with a reduction in grain size, and for the compact with a grain size $< 5\ \mu$m falls to 3.10 W/Km, a reduction of $\sim$28 per cent. At 1000 K the reduction is almost 35 per cent. Measurements of the temperature variation of the transport properties of these materials indicate that the conversion efficiency of a compact with a grain size of $< 5\mu$m would be 17.5 per cent when operating between room temperature and 1000 K, as shown in Figure 5.18. This is 22 per cent higher than the efficiency of a 'single crystal material'.[52]

5.3 MEASUREMENT OF THERMOELECTRIC TRANSPORT PROPERTIES

5.3.1 Introduction

In order to assess the performance of a material intended for use in thermoelectric applications, the figure of merit must be evaluated over the

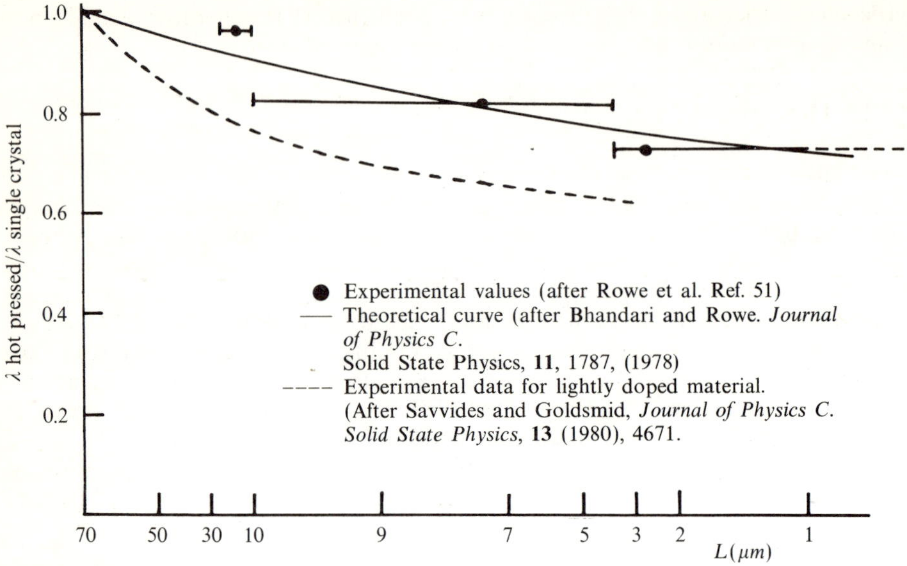

Figure 5.16 $\lambda_{\text{hot pressed}}/\lambda_{\text{single crystal}}$ as a function of grain size for heavily doped silicon–germanium alloys.

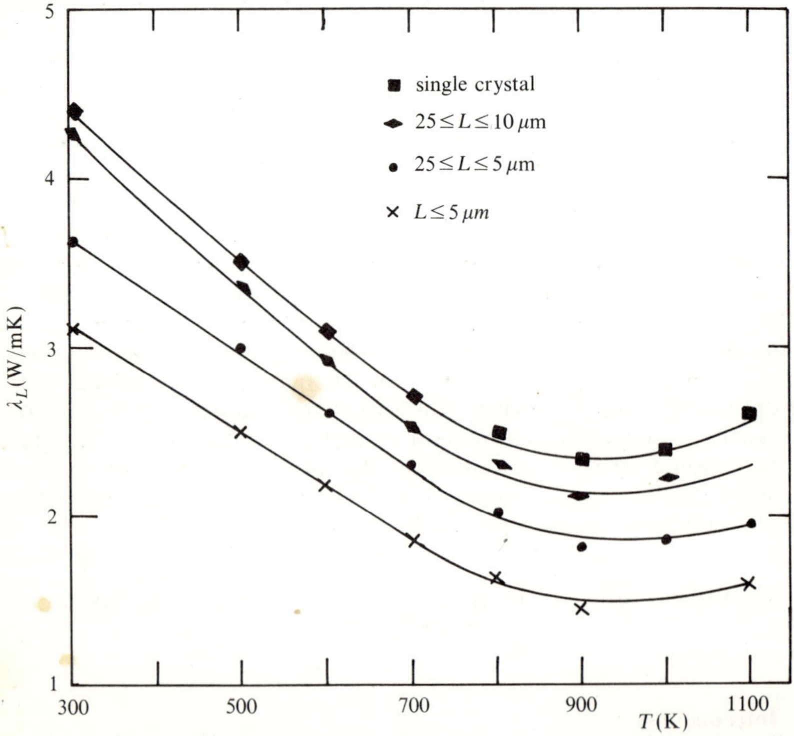

Figure 5.17 Temperature variation of the lattice thermal conductivity of $Si_{63.5}Ge_{36.5}$ alloys for different grain size. (After Rowe and Shukla, reference 52.)

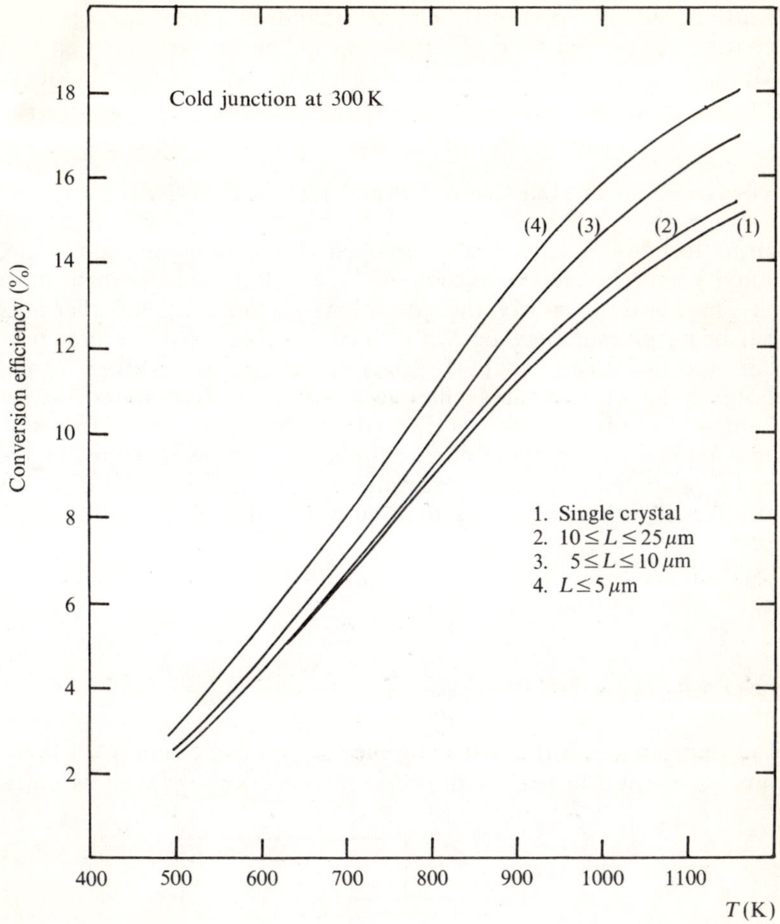

Figure 5.18 Temperature variation of the maximum conversion efficiency η of $Si_{63.5}Ge_{36.5}$ for different grain sizes. (After Rowe and Shukla, reference 52.)

intended range of operation of the device. This involves measurement of Seebeck coefficient, electrical resistivity and thermal conductivity from about room temperature to 1300 K. In addition, since optimization of the figure of merit depends upon the amount of dopant present in the material, information on the active carrier concentration should be obtained from Hall effect measurements. Apart from difficulties in interpreting thermoelectric data on inhomogeneous materials[53] or on anisotropic materials such as bismuth telluride or those prepared by hot pressing,[54] the measurement of transport properties themselves presents a number of problems, especially at high temperature.

The figure of merit optimizes at carrier concentration values of $\sim 10^{25}$ to $10^{26}\ m^{-3}$; consequently the electrical resistivity and Hall voltages will be small ($\sim 10\,\mu\Omega m$ and $\sim 10\,\mu V$) and their measurement requires the use of sensitive measuring techniques. The Seebeck coefficient, on the other hand, will be high (150 to 300 $\mu V/K$) and will give rise to relatively large unwanted voltages if temperature gradients exist in the circuit. In addition, the transport properties may change during the time of measurement, particularly at high temperatures, due to loss of volatile material or a change in the amount of dopant in solid solution.

5.3.2 Measurement of Hall Coefficient and Electrical Resistivity

Numerous techniques have been employed in the measurement of electrical conductivity and the Hall coefficient,[55,56,57,58] and many of them have been specially modified to satisfy the conditions dictated by the thermoelectric material being investigated. Basically there are three main techniques which are employed in measurement: (1) direct current, (2) alternating current, and (3) chopped direct current techniques. All three techniques involve the measurement of voltage, whether it be the voltage drop across the sample on passing a known current as in electrical resistivity measurements, or the Hall voltage V_H generated as a result of applying a magnetic field[58] in the appropriate sample direction. Referring to Figure 5.19, the Hall coefficient (R_H) is given by:

$$R_H = \frac{bV_H}{IB} \quad 5.1$$

and the density of carriers $(n) = \dfrac{r_H}{R_H e}$ 5.2

where r_H depends upon the scattering mechanism (see Section 4.3.4) but to a good approximation in heavily doped semiconductors is taken as unity.

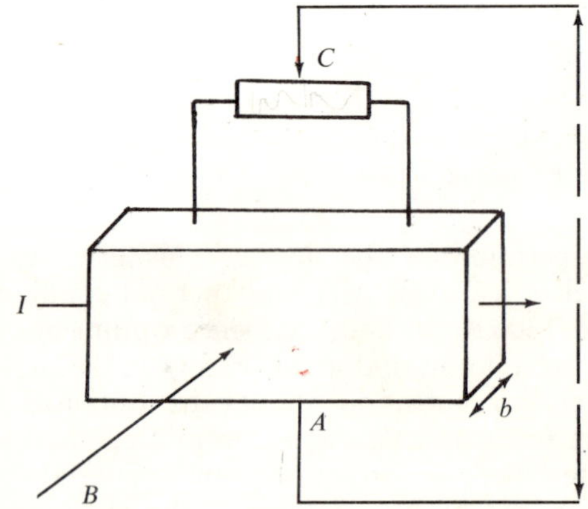

Figure 5.19 Three-probe techniques for Hall effect measurement.

The obvious choice of a method involving the basic measurements of current and potential difference using a d.c. potentiometer is unsuitable for use with thermoelectric materials. The high electrical conductivity of these materials requires the use of relatively large current densities in order to obtain a reasonable potential drop across the specimen. The use of large current densities results in the establishment of a temperature gradient across the sample due to Peltier heating. This is in addition to one which results from asymmetry in the circuit. If no attempt is made to correct for thermoelectric voltages arising from such temperature gradients, serious errors will occur in measurement, particularly in the Hall voltage. Because of the low electrical resistivity of thermoelectric materials it is very difficult to make contacts which are negligible compared to the sample resistance, consequently four-probe techniques must be used in the measurement of electrical resistivity. To ensure that the current flow between the voltage probes is axial the voltage probes should be inset from the current contacts by a distance which is greater than the width of the sample. The error introduced in using direct current is independent of current levels used, and there can be an appreciable buildup of thermal voltage within less than 100 ms of switching on the current.[59] Consequently unwanted voltages arising from asymmetry in the circuit cannot be 'removed' by using a low current level, reversing its direction and taking an average.

A variety of systems which employ alternating current techniques have been developed in an attempt to overcome some of these difficulties. Goldsmid[60] has described a conductivity bridge which is adequate for measurements on bismuth telluride with conductivities in the range 10^4 to $10^6\,\Omega^{-1}\,m^{-1}$, while Dauphinee and Mooser[61] describe a chopped d.c. circuit capable of measuring resistances from 10^{-3} to $10^5\,\Omega$ and Hall coefficients from 10^{-8} to $10^{-1}\,m^3/C$. Thermoelectric materials have a low carrier mobility, and it is often difficult to align the Hall probes sufficiently accurately to avoid large resistance drops between the probes. A three-probe technique[13] can be used to overcome this as shown in Figure 5.19. With the magnetic field off, a null balance is obtained between the probe *A* and the variable 'virtual' probe *C*.

A variety of a.c. methods are available for the measurement of Hall voltage. An alternating current can be supplied to a sample in a static magnetic field[62,63] or a time varying magnetic field can be applied to a sample carrying a d.c. current.[64] Double a.c. techniques have also been employed in which both the sample current and magnetic field vary in time.[65–67]

In principle, chopped d.c. methods should possess the advantages of a.c. while retaining the simplicity associated with the detection of d.c. voltages. A relatively simple system, using a chopped current and a d.c. magnetic field suitable for measurement of electrical resistivity and Hall voltages from room temperature to 1300 K, has been described by Rowe and Bunce.[68] An improved version[69] is shown in Figure 5.20. The function generator provides a 50 per cent duty cycle to the sample. The sample current is amplified to a maximum of 100 mA and a transformer used to isolate the sample circuit from the function generator. The reference signal to the phase sensitive detector is derived from the same function generator and its phase is adjusted to that of the Hall voltage by passing it through a phase changer before presenting it to the phase sensitive detector. The output signal from the sample is fed to a carefully shielded centre tapped transformer with a turns ratio of 16:1. Full wave connection is used because of its larger output and balanced configuration with respect to interfering a.c. voltages. Use of this transformer avoids the need for preamplification and also provides impedance matching which serves to increase the signal-to-noise ratio. The overall sensitivity of the system is critically dependent upon the gain of the preamplifier, but at 70 dB a

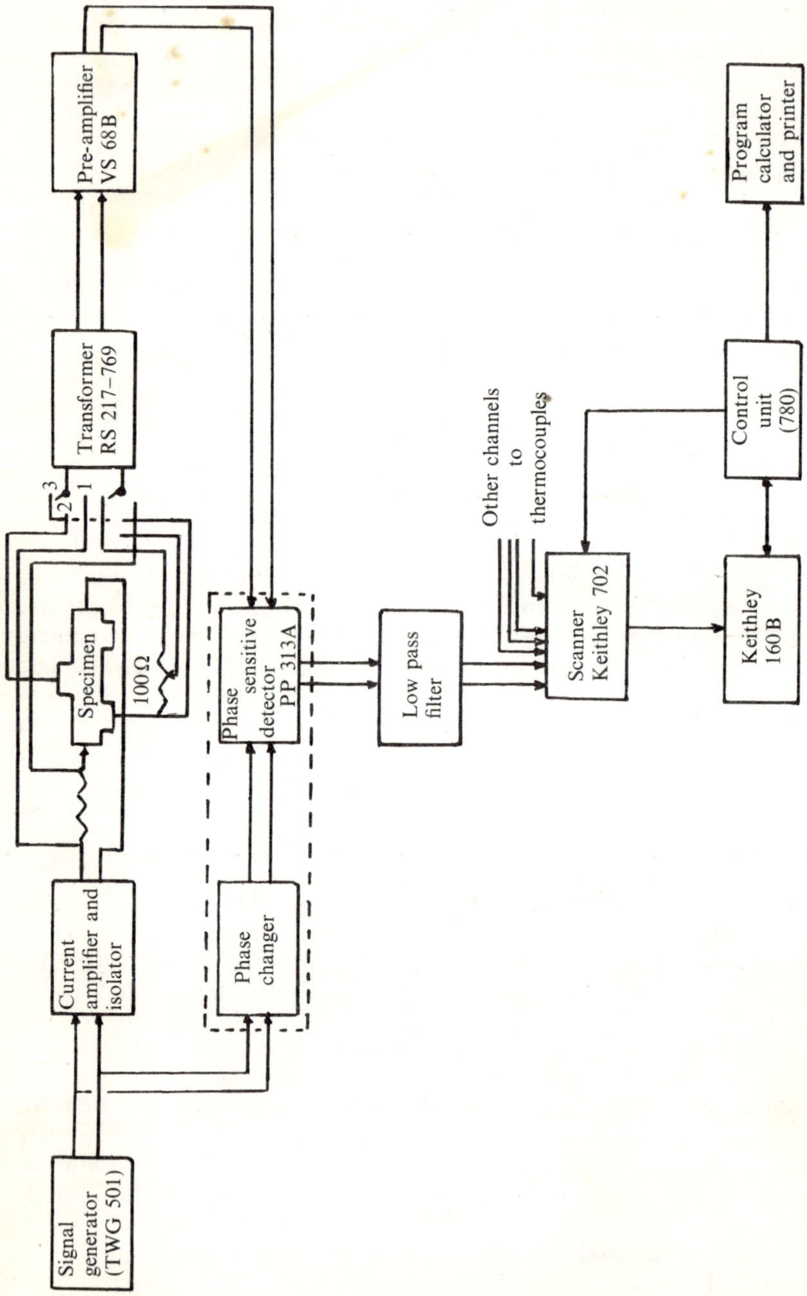

Figure 5.20 Block diagram of electrical circuit for measurement of Hall voltage and electrical resistivity.

sensitivity of 10^{-8} V at room temperature is easily obtained. Reproducibility of measurements is better than ± 2 per cent at 1000 K. With the switch in position 1, the current through the sample is accurately determined from the voltage drop across the standard 1Ω resistance. A three-probe technique is used in detecting the Hall voltage with selection between the resistance and Hall voltage measurement made by switching between positions 1 and 2.

5.3.3 Measurement of Thermal Conductivity

The thermal conductivity λ is defined by the equation:

$$\dot{Q} = -\lambda \nabla T \qquad 5.3$$

where $\dot{Q}$ is the flux of heat maintained in a sample by a temperature gradient ∇T. Thus in order to obtain a value for λ it is necessary to measure a temperature gradient and the corresponding heat flow along the gradient. Thermal conductivity is much more difficult to measure than electrical conductivity, principally due to the fact that thermal insulation is much less effective than electrical insulation and unwanted heat transfer takes place between the sample and the surroundings. In addition the thermal analogue of a high impedance voltage-measuring circuit is difficult to obtain in practice.

Several methods are available for the measurement of thermal conductivity and these are adequately described in standard texts.[70,71] They can be divided into steady and non-steady methods, depending on whether or not the temperature distribution in the sample is time dependent. In steady state methods the temperature gradient obtained under steady conditions when a known heat flux passes through the sample is measured. An inherent drawback of this method is the difficulty of allowing for heat losses, and this applies to thermoelectric materials which are poor conductors of heat; radiant heat losses may be large in comparison with the heat being conducted along the sample. Comparator methods, in which the disc-shaped sample of unknown thermal conductivity is sandwiched between discs of known thermal conductivity and the temperature gradient in the unknown disc and the two standards is then measured, are more suitable for measurements on thermoelectric materials.[72]

Non-steady or dynamic methods enable radiation effects to be eliminated relatively easily, and such methods are widely used in measurements on thermoelectric materials. These techniques generally involve the measurement of thermal diffusivity, α', rather than thermal conductivity, with the two related by:

$$\lambda = \alpha' C'_p \qquad 5.4$$

where C'_p, the heat capacity per unit volume at constant pressure is given by $C_p \rho$; C_p is the specific heat and ρ the density. The temperature dependence of C'_p for hot pressed $Si_{80}Ge_{20}$ alloy is shown in Figure 5.21. Although separate determinations of C_p and ρ are required, their determination to an acceptable accuracy is not difficult. In addition, since they are not 'structure sensitive' it is not necessary to determine them for every sample for which the figure of merit is required. Measurements of thermal diffusivity have been most frequently made using a technique based upon a method developed by

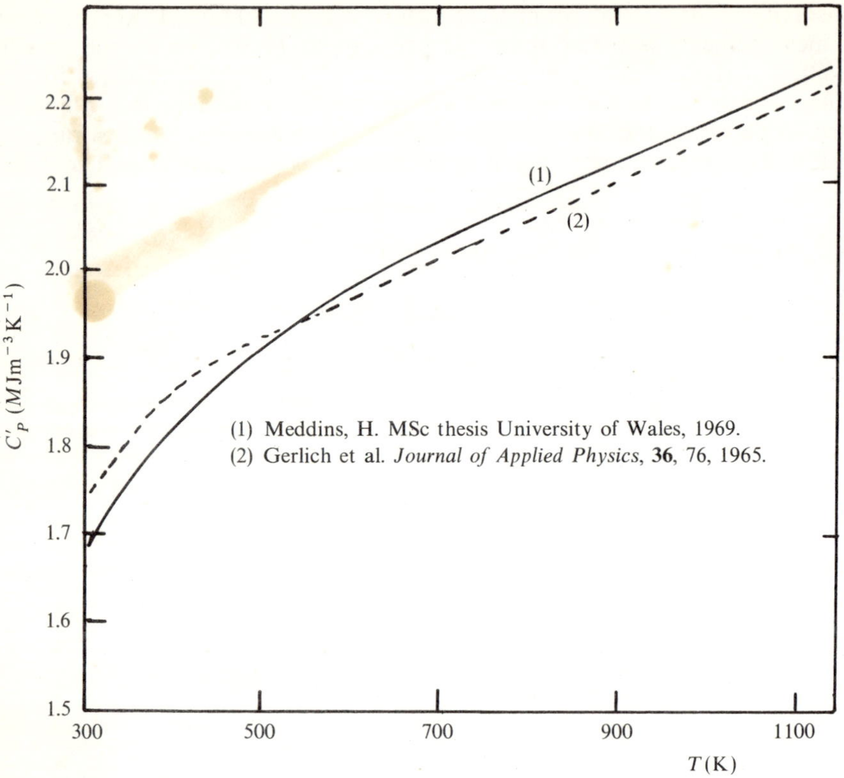

Figure 5.21 Heat capacity per unit volume of $Si_{80}Ge_{20}$ alloy as a function of temperature.

Ångström.[73] In this method a periodic heat input of period $2\pi/\omega$ is supplied to one end of a bar-shaped specimen. The resulting temperature wave is attenuated as it moves along the sample, and:

$$\alpha' = \frac{\omega l^2}{2 \ln \delta_1 . \Phi} \qquad 5.5$$

where δ_1 is the ratio of the amplitudes of the wave at two positions separated by a distance l and the phase difference between these positions is Φ. Novel modifications of this method are generally concerned with the way in which the periodic heat input is generated and presented to the sample.[74,75,76] Although perfectly satisfactory for long rod-shaped samples, the method cannot be used for measurements on short samples or discs. Measurements on small samples with disc-like geometry can be made using the 'flash technique'. In this technique the front face of the sample is subjected to a short burst of radiant energy. The source of energy is usually a xenon flash tube or a laser. The resultant temperature history on the back surface of the sample is recorded and the thermal diffusivity obtained from the relationship:

$$\alpha' = \frac{1.37 a^2}{\pi^2 t_{\frac{1}{2}}} \qquad 5.6$$

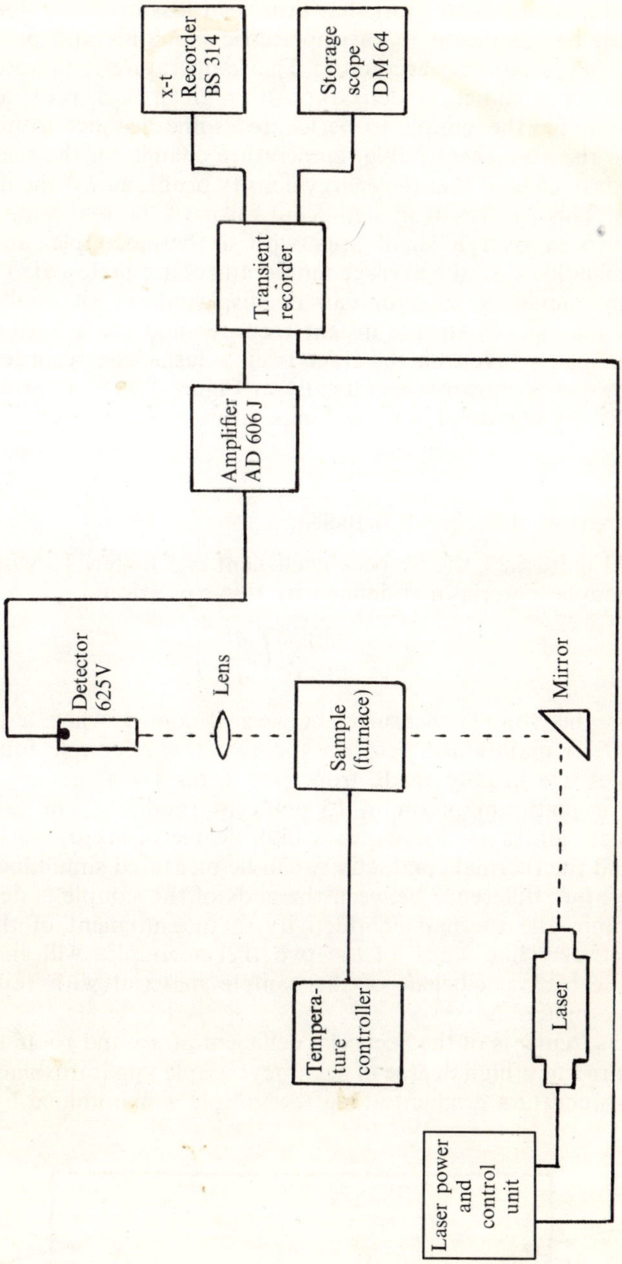

Figure 5.22 Block diagram of thermal diffusivity apparatus. (After Rowe and Shukla, reference 52.)

where a is the sample thickness and $t_{\frac{1}{2}}$ the time taken for the back surface temperature to reach half its maximum value. Although heat losses are minimized by suitable choice of sample thickness and by making measurements in a time which is so short that little heat loss from the sample occurs, they may still be significant in thermoelectric materials and procedures are available to take them into account.[77] The temperature is monitored with a thermocouple or a radiation detector. Although a laser provides sufficient radiation to enable the sample to be located some distance from the source and still give rise to a measurable temperature change on the rear surface, it does have a drawback in that the energy density profile across the laser beam is non-uniform. This can result in significant errors if the rear surface temperature is monitored over a small area with a thermocouple, and radiation detectors which 'look at' the average temperatures are preferred. The response of the thermocouple or detector can be displayed on an oscilloscope and photographed or stored in a transient recorder and the $t_{\frac{1}{2}}$ value computed from measurements taken off the trace from a fast chart recorder. A typical laser flash system is shown schematically in Figure 5.22.[78] Reproducibility of thermal diffusivity measured at room temperature is ± 2 per cent.

5.3.4 Measurement of Seebeck Coefficient

Referring to Figure 5.23, the Seebeck coefficient of a material s with respect to the thermocouple material a is defined by the expression:

$$\alpha_{sa} = \lim_{\Delta T \to 0} \left(\frac{\Delta V}{\Delta T} \right) \qquad 5.7$$

where ΔV is the voltage measured between 1 and 3 when a temperature difference ΔT is maintained between the two thermocouple junctions. The thermocouples are usually made from fine wires ($\sim$40 gauge) of copper–constantan or platinum–platinum/15 per cent rhodium. The Seebeck coefficients of these alloys are known to a high degree of accuracy. The Seebeck coefficient and the thermal conductivity can be measured simultaneously since, if the temperature difference between the ends of the sample is determined in order to obtain the thermal conductivity, a measurement of the potential difference between 'like wires' of the two thermocouples will enable one to obtain the Seebeck coefficient of the sample material with respect to the thermocouple material.

When measurements of the Seebeck coefficient at around room temperature are not required to a high degree of accuracy, simple apparatus may be used in which the temperature gradient along the sample is maintained by sandwich-

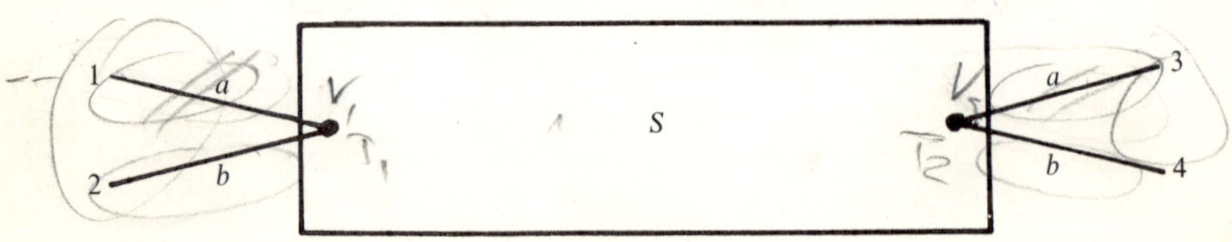

Figure 5.23 Connections for the measurement of Seebeck coefficient.

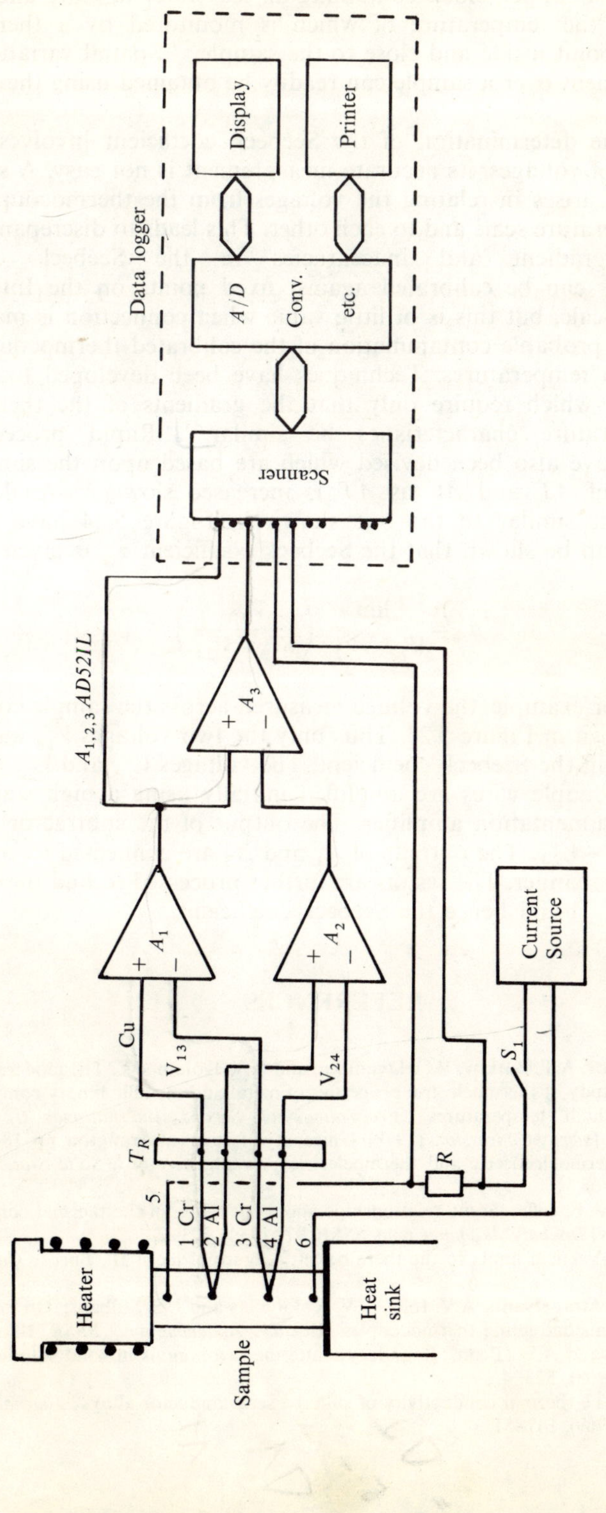

Figure 5.24 Analogue subtraction circuit for the measurement of Seebeck coefficient. (After Shukla, reference 78.)

ing it between a copper block containing an ice–water mixture and a heated copper block, the temperature of which is monitored by a thermocouple soldered at a point inside and close to the sample.[79] Spatial variations of the Seebeck coefficient over a sample can readily be obtained using the hot probe technique.[80]

Although the determination of the Seebeck coefficient involves only the measurement of voltages, its accurate measurement is not easy. A substantial difficulty often arises in relating the voltages from the thermocouples to the absolute temperature scale and to each other. This leads to discrepancies in the temperature gradient and inaccuracies in the Seebeck coefficient. Thermocouples can be calibrated against fixed points on the International Temperature Scale, but this is of little value when connection is made to the sample due to probable contamination of the calibrated thermocouple, particularly at high temperatures. Techniques have been developed to overcome this drawback which require only that the gradients of the thermocouple voltage–temperature characteristics be similar.[81] Rapid procedures for evaluating α have also been devised which are based upon the simultaneous measurement of ΔT and ΔV as ΔT is increased slowly.[82] Analogue subtraction circuits similar to the one shown in Figure 5.24 have also been employed. It can be shown that the Seebeck coefficient α_{sa} is given by:[83]

$$\alpha_{sa} = \lim_{V_{13} \to V_{24}} \left(\frac{V_{13}}{V_{13} - V_{24}} \right) \tag{5.8}$$

where V_{13} is, for example, the voltage measured across the sample connections 1 and 3 as shown in Figure 5.23. Thus only the two voltages V_{13} and V_{24} are needed to obtain the Seebeck coefficient. The voltages V_{13} and V_{24} across the similar thermocouple wires are amplified initially using a high stability and low noise instrumentation amplifier. The output of the subtractor A_3 yields the voltage $V_{13} - V_{24}$. The outputs of A_1 and A_3 are connected to consecutive channels of the scanner. The results are further processed to find the best slope of $V_{13}/(V_{13} - V_{24})$ and hence the Seebeck coefficient.

REFERENCES

1. M.V. Vedernikov, A.T. Burkov, V.G. Dvunitkin and N.V. Kolomoets, 'The modern state of the experimental study of thermoelectric properties of metal and metallic binary continuous solid solutions at high temperatures', *Proceedings of Second International Conference on Thermoelectric Energy Conversion*, (1978), University of Texas at Arlington, pp. 18–23.
2. F.D. Rosi, 'Thermoelectricity and thermoelectric power generators', *Solid State Electronics*, **11** (1968), 833–68.
3. A.V. Ioffe and A.F. Ioffe, 'Some relationships about the value of the thermal conductivity of semiconductors', *Dokladȳ Akademie nauk SSSR*, **97** (1954), 821–2.
4. R.W. Ure Jr, 'Practical limits to the thermoelectric figure of merit II', *Energy Conversion*, **12** (1971), 45–52.
5. A.F. Ioffe, S.V. Airapetyants, A.V. Ioffe, N.V. Kolomoets and L.S. Stilbans, 'On increasing the efficiency of semiconducting thermocouples', *Dokladȳ Akademie nauk SSSR*, **106** (1956), 981.
6. H.J. Goldsmid and A.W. Penn, 'Boundary scattering of phonons in solid solutions', *Physics Letters*, **27A** (1968), 523–4.
7. J.E. Parrott, 'The thermal conductivity of sintered semiconductor alloys', *Journal of Physical Chemistry*, **2** (1969), 147–51.

8. H.R. Meddins and J.E. Parrott, 'The thermal and thermoelectric properties of sintered germanium–silicon alloys', *Journal of Physics C*, **9** (1976), 1263–76.
9. C.M. Bhandari and D.M. Rowe, 'Thermal conductivity of highly disordered semiconductor alloys', *Journal of Physics D; Applied Physics* **10** (1977), 59–61.
10. D.M. Rowe and C.M. Bhandari, 'Fine grained Si–Ge alloys as superior thermoelectric materials', *Proceedings of Second International Conference on Thermoelectric Energy Conversion*, (1978), University of Texas at Arlington, pp. 32–6.
11. R.K. Pisharody and L.P. Garvey, 'Modified silicon–germanium alloys with improved performance', *Proceedings of 13th Intersociety Energy Conversion Engineering Conference* (*IECEC*), (1978), San Diego, California, Vol. 3, pp. 1963–8.
12. D. Tuomi, 'Thermoelectric alloys. Phase diagrams and imperfection chemistry', *Proceedings of the Third International Conference on Thermoelectric Energy Conversion*, (1980), University of Texas at Arlington, pp. 174–83.
13. R.R. Heikes and R.W. Ure, *Thermoelectricity; Science and Engineering*, Interscience, London, (1961), Chapters 11 and 13.
14. M.W. Nicolaou, 'Materials for direct thermoelectric conversion with a high figure of merit', *Proceedings of the First International Conference on Thermoelectric Energy Conversion*, (1976), University of Texas at Arlington, pp. 59–63.
15. D.A. Wright, 'Materials for direct conversion thermoelectric generators', *Metallurgical Reviews*, **15** (1970), 147–60.
16. J.R. Drabble, R.D. Groves and R. Wolfe, 'Galvanomagnetic effects in *n*-type bismuth telluride', *Proceedings of the Physical Society*, **71** (1958), 430–43.
17. I.G. Austin, 'The optical properties of bismuth telluride', *Proceedings of the Physical Society*, **72** (1958), 545–52.
18. H.J. Goldsmid, A. R. Sheard and D. A. Wright, 'The performance of bismuth telluride thermojunctions', *British Journal of Applied Physics*, **9** (1958), 365–70.
19. W.D. Lawson, *Preparation of Single Crystals*, Academic Press, London, 1958.
20. F.D. Rosi, E.F. Hocking and N.E. Lindenblad, 'Semiconducting materials for thermoelectric power generation', *RCA Review*, **22** (1961), 82–121.
21. J.W. McGrew, 'A report on the properties and performance of TAGS', *Proceedings of Fifth IECEC*, (1970), Las Vegas, Nevada, 1531–40.
22. B. Abeles, D.S. Beers, G.D. Cody and J.P. Dismukes, 'Thermal conductivity of Ge–Si alloys at high temperatures' *Physical Review*, **125** (1962), 44–6.
23. B. Abeles and R.W. Cohen, 'Ge–Si thermoelectric power generator', *Journal of Applied Physics*, **35** (1964), 247–8.
24. B. Abeles, 'Lattice thermal conductivity of disordered semiconductor alloys at high temperatures', *Physical Review*, **131** (1963), 1906–11.
25. W.G. Pfann, '*Liquid Metals and Solidification*', American Society for Metals, Cleveland, Ohio, (1958), p. 233.
26. J.P. Dismukes and L. Ekstrom, 'Homogeneous solidification of Ge–Si alloys', *Transactions of the Metallurgical Society* AIME **233** (1965), 672–80.
27. W.H. Dietz and H.A. Herrmann, 'Methods of preparation of Ge–Si alloys', *Kristall und Technik*, **4** (1969), 413–21.
28. F.A. Trumbore, 'Solid solubilities of impurity elements in germanium and silicon', *Bell System Technical Journal*, **39** (1960), 205–33.
29. L. Ekstrom and J.P. Dismukes, 'Precipitation of phosphorus from solid solution in Ge–Si alloy, *Journal of Physics and Chemistry of Solids*, (1966), 857–63.
30. N. Savvides and D.M. Rowe, 'Precipitation of phosphorus from solid solutions of Si–Ge alloys and its effect on thermoelectric transport properties', *Journal of Physics D*, **14** (1981), 723–32.
31. V.S. Shukla and D.M. Rowe, 'Precipitation of boron from silicon–germanium alloy and its effect on the thermoelectric transport properties', *Physica Status Solidi*, (a) **66** (1981), 243–8.
32. V. Raag, 'Comprehensive thermoelectric properties of *n*- and *p*-type 78% Si–22% Ge', *Proceedings of the Second International Conference on Thermoelectric Energy Conversion*, (March 1978), University of Texas at Arlington, pp. 5–10.
33. W.C. Mitchell et al, 'Transport theory of 3M high performance thermoelectric materials', *Proceedings of the 11th IECEC*, (September 1976), State Line, Nevada, pp. 1539–45.
34. G. Stapfer and L.P. Garvey, 'Selenide technology evaluation program at Jet Propulsion Laboratory', *Proceedings of the 13th IECEC*, (August 1978), San Diego, California, pp. 1946–51.

35. J.D. Hinderman, P.B. Hogerton and R. Schoenkerr, 'Analytical model for predicting performance and degradation of selenide thermoelectrics', *Proceedings of Third International Conference on Thermoelectric Energy Conversion*, (March 1980), University of Texas at Arlington, pp. 13–20.
36. E.F. Hampl Jr, R.B. Ericson, W.C. Mitchell, R.S. Reylek, 'TPM-217 material performance characteristics', *Proceedings of the 7th IECEC*, (September 1972), San Diego, California, Paper No. 729040.
37. J.C. Bass and N.B. Elsner, 'Segmented selenide thermoelectric generator', *Proceedings of the Third International Conference on Thermoelectric Energy Conversion*, (March 1980), University of Texas at Arlington, pp. 8–12.
38. E.F. Hampl et al., 'Design of a thermoelectric converter using 3M high performance thermoelectric material', *Proceedings of the 10th IECEC*, (August 1975), Newark, Delaware, pp. 714–22.
39. A.R. Lieberman, W.E. Osmeyer and T.E. Hammel, 'The Low Cost High Performance Generator (LCHG)', *Proceedings of the 11th IECEC*, (September 1976), State Line, Nevada, pp. 1591–8.
40. G. Stapfer, 'The thermoelectric technology program at the Jet Propulsion Laboratory', *Proceedings of Third International Conference on Thermoelectric Energy Conversion*, (March 1980), University of Texas at Arlington, pp. 70–5.
41. A.R. Lieberman, (1981). Private communication.
42. P.A. O'Riordan, 'The US Department of Energy's thermoelectric development program for space power' *Proceedings of 4th ICTEC* (March 1982) University of Texas at Arlington, pp. 15–25.
43. G.L. McVay, R.A. Lefever and R.J. Baughman, 'Preparation of hot pressed silicon–germanium ingots, II. Reduction of chill cast-material', *Materials Research Bulletin* **9**, (1974), 735–44.
44. H. Schreiner and F. Wendler, 'Die Theorien der Sintering im Hinblick auf Effekte bei der pulvermetallurgischen Herstellung thermoelektrischer Werkstoffe', *Zeitschrift für Metallkonde*, **52** (1961), 218–23.
45. W.R. George, R. Sharples and J.E. Thompson, 'The sintering of bismuth telluride', *Proceedings of the Physical Society*, **74** (1959), 768–70.
46. M. Piper, 'Germanium–silicon thermoelectric material development', *Proceedings of UKAEA-ENEA, International Symposium on Industrial Applications for Isotopic Power Generators*, (1966), Harwell, UKAEA, pp. 349–61.
47. R.A. Lefever, G.L. McVay and R.J. Baughman, 'Preparation of hot-pressed silicon–germanium ingots. III. Vacuum hot pressing', *Material Research Bulletin*, **9** (1974), 863–72.
48. D.M. Rowe and R.W. Bunce, 'The vacuum hot pressing of germanium and silicon–germanium alloys', *Journal of Physics D*, **10** (1977), 941–8.
49. N. Savvides and H.J. Goldsmid, 'Hot-press sintering of Ge–Si alloys', *Journal of Materials Science*, **15** (1980), 594–600.
50. G.H. Reynolds, N. B. Elsner and P.H. Miller Jr, 'Development of thermoelectric materials for operation at 1200–1500°C', *Proceedings of Third International Conference on Thermoelectric Energy Conversion*, (March 1980), University of Texas at Arlington, pp. 67–9.
51. D.M. Rowe, V.S. Shukla and N. Savvides, 'Phonon scattering at grain boundaries in heavily-doped, fine grained silicon–germanium alloys', *Nature* **290** (1981), 5809, 765–766.
52. D.M. Rowe and V.S. Shukla, 'The effect of grain boundary scattering on the lattice thermal conductivity and thermoelectric conversion efficiency of heavily doped, fine grained silicon–germanium alloys', *Journal of Applied Physics*, **52** (1981), 12, 7421–6.
53. C. Herring, 'Effect of random inhomogeneities on electrical and galvanomagnetic measurements', *Journal of Applied Physics*, **31** (1960), 11, 1939–53.
54. D. Ryden, 'A comparison between the transport properties of single crystal, polycrystal and powder anisotropic semiconductors', *Solid State Physics*, **4** (1971), 1192–205.
55. J.P. Jan, *Solid State Physics*, (Ed.) F. Seitz and D. Turnbull, Academic Press, London, Vol. 5, 1957.
56. L. Marton, *Methods of Experimental Physics*, Academic Press, London, Vol. 6B, 1959.
57. E.H. Putley, *The Hall Effect and Related Phenomena*, Butterworth, London, 1960.
58. H.H. Weider, *Material Science Monograph*, Elsevier, New York, 1979.
59. D. Ryden, 'Techniques for the measurement of the semiconductor properties of thermoelectric materials', *AERE Report R*, 6996, (1973).
60. H.J. Goldsmid, PhD Thesis, University of London, 1957.
61. T.M. Dauphinee and E. Mooser, 'Apparatus for measuring resistivity and Hall coefficient of semiconductors', *Review of Scientific Instruments*, **26** (1955), 660–4.

62. J.J. Donoghue and W.P. Eatherly, 'A new method for precision measurements of the Hall and magneto-resistance coefficients', *Review of Scientific Instruments*, **22** (1951), 513–20.
63. J. Lavine, 'Alternating current apparatus for measuring the ordinary Hall coefficient of ferromagnetic metals and semiconductors', *Review of Scientific Instruments*, **29** (1958), 970–6.
64. R.A. Lomas, M.J. Hampshire and R.D. Tomlinson, 'A sensitive method of Hall measurement', *Journal of Physics E: Scientific Instruments*, **5** (1972), 819–21.
65. B.R. Russel and C. Wahling, 'A new method for the measurement of Hall coefficients', *Review of Scientific Instruments*, **21** (1950), 1028–9.
66. N.Z. Lupu, N.M. Tallan and D.S. Tannhauser, 'Apparatus for measuring the Hall effect of low mobility samples at high temperatures', *Review of Scientifie Instruments*, **38**, (1967), 1658–61.
67. H.L. McKinzie and D.S. Tannhauser, 'Systematic errors in alternating current Hall effect measurements', *Journal of Applied Physics*, **40** (1969), 4954–8.
68. D.M. Rowe and R.W. Bunce, 'Apparatus for measuring resistivity and Hall coefficient of heavily doped semiconductors at high temperatures', *Journal of Physics E: Scientific Instruments*, **4** (1971), 902–4.
69. V.S. Shukla and D.M. Rowe, 'Effect of short-term heat treatment on the thermoelectric properties of heavily doped n-type silicon–germanium alloys', *Applied Energy*, **9** (1981), 131–9.
70. J.R. Drabble and H.J. Goldsmid, *Thermal Conduction in Semiconductors*, Pergamon Press, Oxford, 1962.
71. J.E. Parrott and A.D. Stuckes, *Thermal Conductivity of Solids*, Pion, London, 1975.
72. A.D. Stuckes and R.P. Chasmar, 'Semiconductor Conference Report', *Physical Society London*, **119** (1956).
73. A.J. Ångström, 'New method of determining the thermal conductivity of bodies', *Philosophical Magazine*, **25** (1863), 130–42.
74. P.H. Sidles and G.C. Danielson, 'Thermal diffusivity of metals at high temperatures', *Journal of Applied Physics*, **25** (1954), 58–66.
75. B. Abeles, G.D. Cody and D.S. Beers, 'Apparatus for the measurement of the thermal diffusivity of solids at high temperatures', *Journal of Applied Physics*, **31** (1960), 1585–92.
76. H.R. Meddins and J.E. Parrott, 'An apparatus for the high temperature measurement of thermal diffusivity, electrical conductivity and Seebeck coefficient', *British Journal of Applied Physics, Journal of Physics D*, (1969), 691–7.
77. R.D. Cowan, 'Pulse method of measuring thermal diffusivity at high temperatures', *Journal of Applied Physics*, **34** (1963), 926–7.
78. V.S. Shukla, 'Evaluation of the thermoelectric performance of improved heavily doped silicon–germanium alloys', PhD Thesis, University of Wales, 1982.
79. See H.J. Goldsmid, *Applications of Thermoelectricity*, Methuen, London, 1960.
80. L.E. Cowles and L.A. Dauncey, 'Apparatus for the rapid scanning of the Seebeck coefficient of semiconductors', *Journal of Scientific Instruments*, **39** (1962), 16–18.
81. D.M. Rowe and R.W. Bunce, 'The thermoelectric properties of heavily doped hot pressed germanium–silicon alloys', *British Journal of Applied Physics* (*Journal of Physics D*), **2**(2) (1969), 1497–502.
82. G.R. Caskey, D.J. Sellmyer and L.G. Rubin, 'A technique for the rapid measurement of thermoelectric power', *Review of Scientific Instruments* (*USA*), **40** (1969), 1280–2.
83. J.E. Ivory, 'Rapid method for measuring Seebeck coefficient as ΔT approaches zero', *Review of Scientific Instruments* (*USA*), **33** (1962), 992–3.

[illegible], M.F. Hampton, and [illegible] 'A simple method of Hall measurement [illegible]' [illegible] 819 [illegible]

[illegible]

6

Thermocouple Construction and Configurations

6.1 THERMOCOUPLE CONSTRUCTION

Two dissimilar arms of *p*- and *n*-type semiconductor material can be joined to form a thermocouple by a variety of methods. The method most frequently used is to bond them together at the hot junction with a strap of good electrical and thermal conductivity and to ensure that the hot source is in good thermal contact with, but electrically insulated from, the thermocouple. Thermocouple junctions can also be made by plastic deformation of the material, with heat and pressure being simultaneously applied to the junction area; hot pressing techniques are also widely employed.

A diagram of a two-couple single-stage thermoelectric generator is shown in Figure 6.1. For high generating efficiency, the device should have a large Z value and operate over as wide a temperature range as possible. However, the figure of merit values are dependent upon the level of doping, with the optimum value increasing with increase in temperature. Consequently, optimum performance can only be attained over a limited temperature range. (The selenides are an exception: see section 5.2.2.)

Ideally the legs of the thermocouples should possess a gradation in doping, but this has not so far been practised. Emphasis has been placed on developing materials which exhibit a high value of Z over a relatively narrow temperature range, and relying on device technology to effectively extend the working temperature range. An approximation to the graded thermocouple is the so-called sandwich arrangement shown in Figure 6.2(a). Each section operates at its best average Z value, and the overall figure of merit is given by the envelope, Figure 6.2(b). The main disadvantage is the necessity of employing additional contacts with their associated resistances. In operation over large temperature differences, the idea can be extended to employing a number of different materials in a segmented arrangement, as shown in Figure 6.3. Practical examples of segmented arrangements are PbTe to SiGe,[1] 80/20 SiGe (hot side) to 63/37 SiGe (cold side)[2]. *N*-type legs have been made by segmenting TEGS 2P and 3P–PbTe material (cold side) with SiGe (hot side).[3] Segmented P-type legs have also been made of BiSbTe (cold side) to 3P (PbSnMnTe) (hot side).[4] Selenide material has also been segmented;

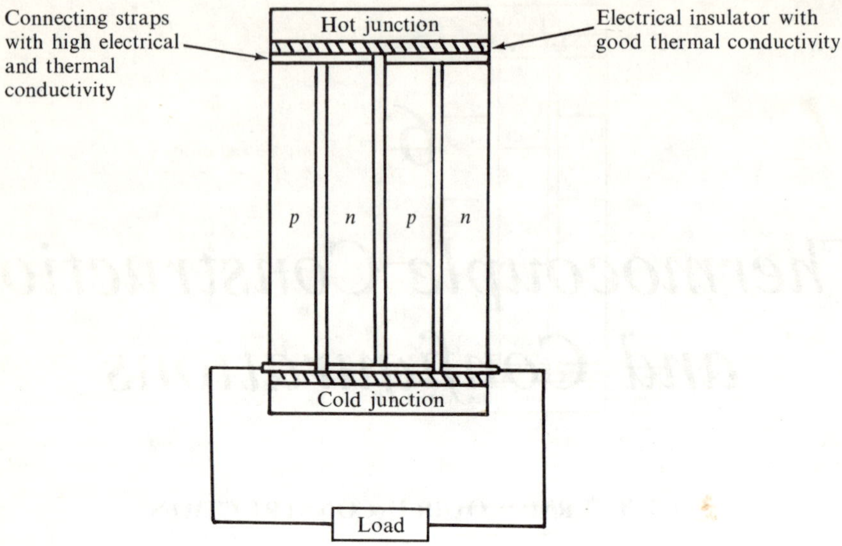

Figure 6.1 Schematic two-couple single-stage generator.

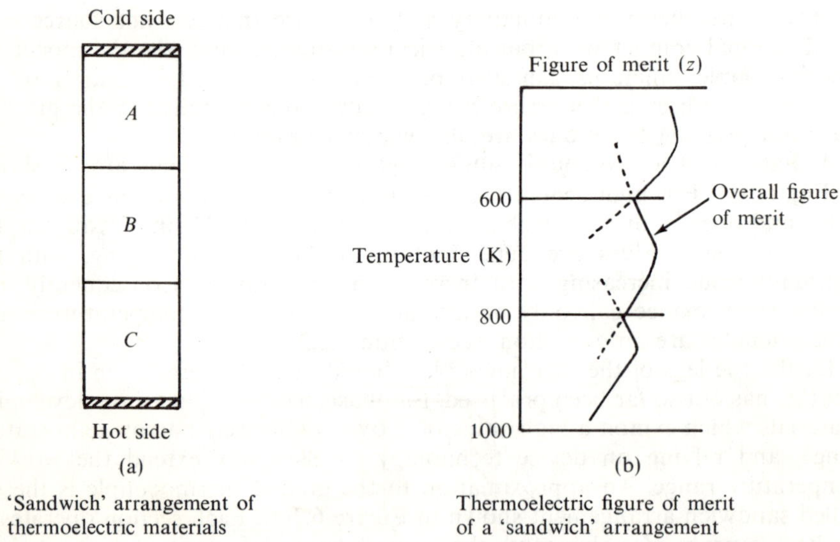

Figure 6.2 (a) Sandwich arrangement of thermoelectric materials. (b) Thermoelectric figure of merit of a sandwich arrangement.

$(CuAg)_2Se$ to $(BiSb)_2Te_3$ forms the *p*-leg and $GdSe_{1.5}$ to PbTe forms the *n*-leg.[5]

Segmenting, however, introduces additional difficulties in satisfying the geometrical requirement of each segment for optimum performance. These difficulties are partially overcome in the cascade arrangement. In this configuration, couples fabricated from different materials are arranged thermally

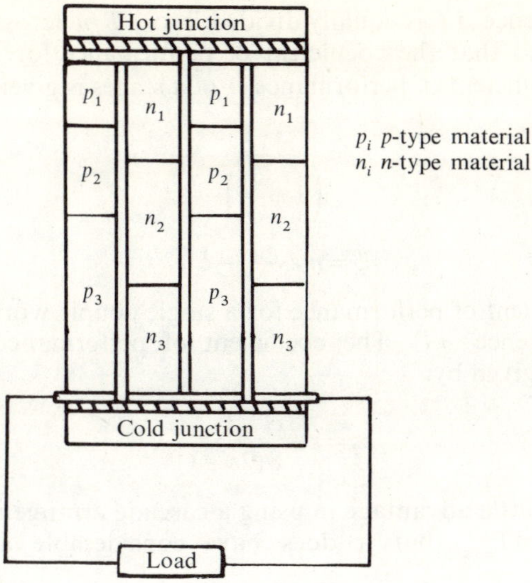

Figure 6.3 Segmented arrangement of thermoelectric materials.

in series and electrically insulated from one another. The second stage of thermocouples uses heat rejected from the first stage and so on, and in practical applications thermal interconnection can be achieved with a heat pipe.[6] However, with two separate loads it suffers from the disadvantage that the electrical conductors connected to the upper thermoelement form a thermal shunt to the lower thermoelement if both loads are at room temperature.

Cascade arrangements are also used in thermoelectric refrigeration to obtain temperature differences greater than the value given by equation 2.22. A schematic two-stage cascade is shown in Figure 6.4. If it is assumed that the

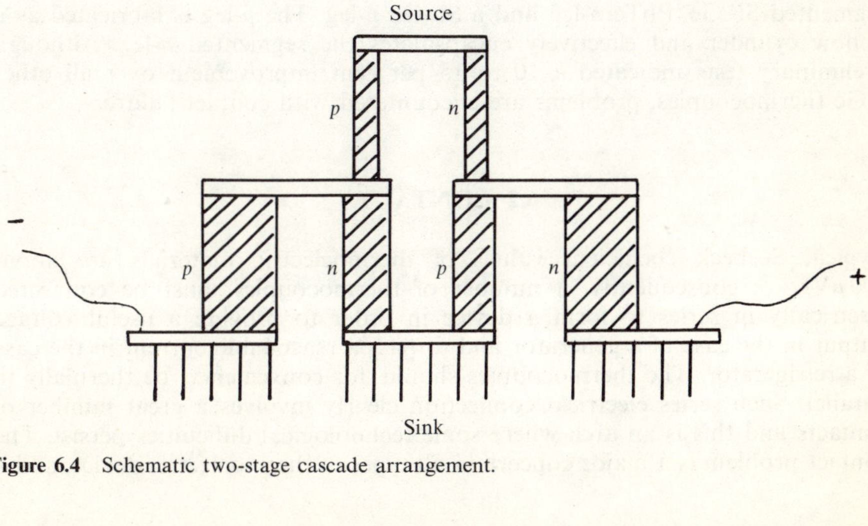

Figure 6.4 Schematic two-stage cascade arrangement.

temperature difference ΔT is equally divided into $\Delta T/n$ across each stage of an n stage device, and that the coefficient of performance for each stage is the same, then the coefficient of performance ε_n of n stages is given by:[7]

$$\varepsilon_n = \frac{1}{\left(1+\frac{1}{\phi}\right)^n - 1} \tag{6.1}$$

where

$$\phi = n(\varepsilon + \tfrac{1}{2}) - \tfrac{1}{2} \tag{6.2}$$

and ε is the coefficient of performance for a single couple working over the full temperature difference ΔT. The coefficient of performance of a two-stage device, ε_2, is then given by:

$$\varepsilon_2 = \varepsilon + \frac{1}{8(2\varepsilon+1)} \tag{6.3}$$

Evidently there is little advantage in using a cascade arrangement if ΔT is very much less than ΔT_{max}, but it does have considerable advantages when $\Delta T \sim \Delta T_{max}$.

As we have noted previously, degradation in material performance accompanies operation at elevated temperatures.[8] In thermoelectric generators problems associated with material sublimation restrict the upper operating temperature of lead telluride to about 800 K[9] unless steps are taken to reduce the free volume available to the lead telluride vapour by sleeving[10] the thermocouple or by encapsulation. Similar problems are encountered in TAGS at temperatures around 800 K. The problem is minimized through the use of microquartz packing or ceramic coatings.[11,12] In silicon–germanium alloys, difficulties are experienced in maintaining thermocouple performance owing to dopant precipitation at high temperature (see section 5.2.2), and ion plating with Si_3N_4 has been employed to suppress vaporization and maintain electrical insulation.[13] Cover gases such as xenon have also been employed to prevent oxidation problems.[14]

The so-called hybrid-thermocouple arrangement[15] overcomes these problems to some extent, and is shown schematically in Figure 6.5. It consists of a segmented Si–Ge/PbTe n-leg and a Si–Ge p-leg. The p-leg is fabricated as a hollow cylinder and effectively encapsulates the segmented n-leg. Although preliminary tests indicated a 10 to 15 per cent improvement over all other SiGe thermocouples, problems are encountered with contact failure.

6.2 CONTACTS

Typical Seebeck coefficient values of thermoelectric materials are about 200 μVK^{-1}; consequently, a number of thermocouples must be connected electrically in series to form a device in order to provide a useful voltage output in the case of a generator and to pass a reasonable current in the case of a refrigerator. The thermocouples should, for convenience, be thermally in parallel. Such series electrical connection clearly involves a great number of contacts and this is an area where some technological difficulties persist. The contact problem is a major concern in all types of thermoelectric device, and a

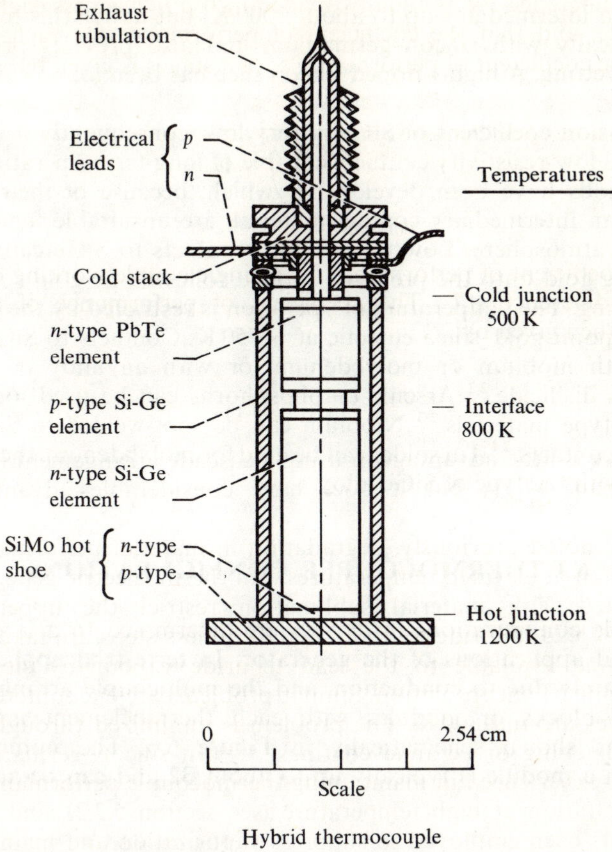

Figure 6.5 Hybrid thermocouple arrangement. (After Bifano, reference 15.)

variety of joining methods have been developed.[16] Bi_2Te_3 and its alloys, at room temperature, do not present insuperable problems, and low electrical resistance contacts can be made by first etching the contact area[17] and then soldering,[7] or by bonding using a pulse technique.[18] In generators care must be taken in making high temperature contacts to avoid the use of any contact material that could affect the thermoelectric properties of the elements. Contacts to Bi_2Te_3, PbTe, SnTe, GeTe and their alloys can be made by pressing iron, nickel, or stainless steel wires into the semiconductor when it is just molten. Contact to PbTe, SnTe and GeTe for operation up to 775 K can be made by diffusing a tin solder into the connecting wires and the semiconductor, and then soldering.[19] Copper can be used as a contact material to Bi_2Te_3, provided that a suitable diffusion barrier is established between the copper and the Bi_2Te_3. It is unsuitable as an intermediary with silicon–germanium as it forms a low melting-point copper–silicon alloy.[19]

The more refractory the material, the more difficult it is to secure a good contact. Gold or aluminium are unsuitable for use with high-temperature thermoelements because both form low-melting-point eutectics. Care must be taken not to use materials that diffuse readily, such as Ge, Ni, Pt. Tungsten is suitable as an intermediary up to about 1000 K, but above this temperature it reacts chemically with silicon–germanium and also presents problems with incomplete wetting. A highly doped silicon slice has been found suitable for the hot junction.[20]

The expansion coefficient of SiGe is very low; consequently it is difficult to make reliable low-resistivity contacts capable of long-term operation. Complex joining methods have been developed[21] which, because of their reliance on tungsten as an intermediary contact material, are unsuitable for operation in an oxidizing atmosphere. Low temperature contacts to SiGe can be made by electroplating gold onto the prepared surface, followed by gold sputtering and pressure baking. The temperature of operation is restricted by the formation of low-melting-point gold–silica eutectic at $\sim$750 K. Contacts to SiGe alloys can be made with niobium or molybdenum, or with an alloy of silicon with molybdenum disilicide.[22] Arsenic or phosphorus can be used for *n*-type and boron for *p*-type materials.[23] Niobium can be spot-welded to SiGe for high temperature contacts.[24] Tin solder can be used for molybdenum, the solder being compatible with the type of SiGe alloy.

6.3 THERMOCOUPLE CONFIGURATIONS

Thermocouple configurations in devices are determined to a large extent by the individual applications of the generator. In terrestrial applications, heat losses are mainly due to conduction, and the multicouple arrangement takes the form of blocks or modules with each thermoelement interconnected electrically as shown schematically in Figure 6.6. The number of thermocouples in a module is typically up to about 32 and can be an even or an

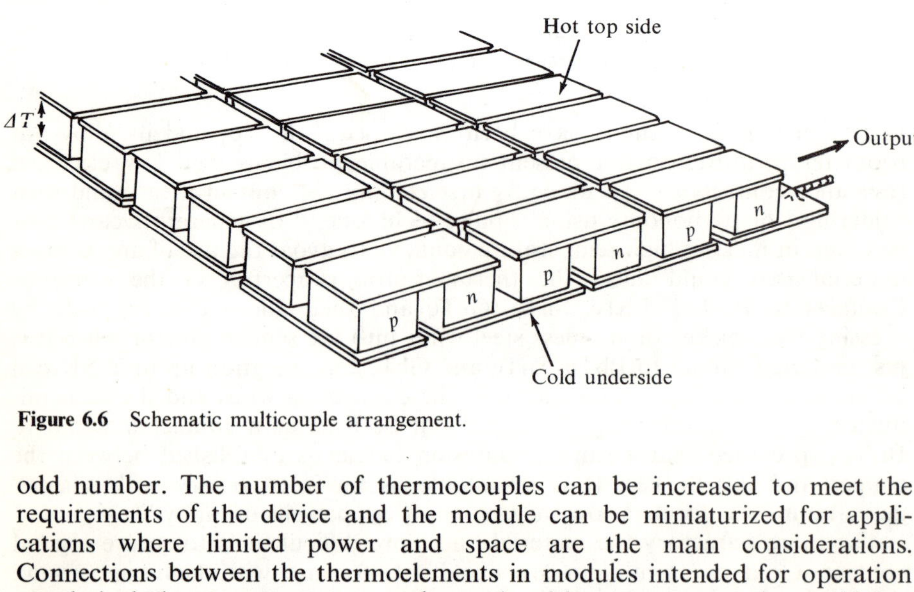

Figure 6.6 Schematic multicouple arrangement.

odd number. The number of thermocouples can be increased to meet the requirements of the device and the module can be miniaturized for applications where limited power and space are the main considerations. Connections between the thermoelements in modules intended for operation at relatively low temperatures can be made with copper straps, but at higher temperatures this technique is unsuitable for use with materials other than

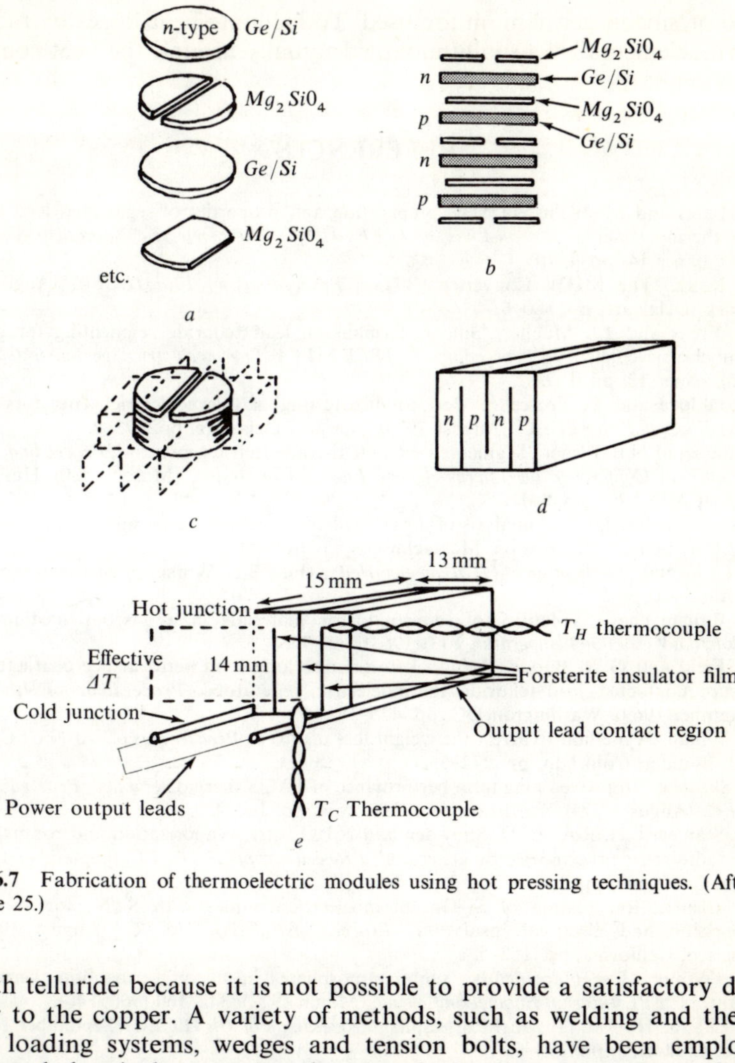

Figure 6.7 Fabrication of thermoelectric modules using hot pressing techniques. (After Piper, reference 25.)

bismuth telluride because it is not possible to provide a satisfactory diffusion barrier to the copper. A variety of methods, such as welding and the use of spring loading systems, wedges and tension bolts, have been employed to make good electrical connections between thermoelements at the module's hot side using interconnecting 'shoes' made from material which is unreactive to the thermoelement material.

Hot pressing techniques greatly facilitate the fabrication of thermoelectric modules as shown in Figure 6.7.[25] In this case cold compacted discs of *n*- and *p*-type Si–Ge alloys are interlaced with sheets of electrically insulating material, and the whole stack is pressed. Although the junctions themselves form an integral part of the structure, problems still exist in making external connections to the module. This problem can be overcome as in the case of segmented PbSnMnTe/BiSbTe elements by including an iron 'cap' in the prepressed stack. Connections can then be made to this cap with laser welding.[26]

In space applications, where heat losses are mainly radiative, cold junction temperatures are relatively high and high-melting-point materials such as lead telluride or silicon germanium are used. To facilitate heat losses by radiation, the thermoelements are usually mounted radially around the heat source.

REFERENCES

1. H.E. Bates and M. Weinstein, 'The preparation and properties of segmented lead telluride–SiGe thermoelements', *Proceedings of IEEE/AIAA. Thermoelectric Specialists Conference*, (1966), paper 14, pp. 1–10.
2. C.E. Kelly, 'The MHW Converter (RTG)', *Proceedings of 10th IECEC*, (August 1975), Newark, Delaware, pp. 880–6.
3. D.G. Freas and J.J. Mueller, 'Silicon–germanium, lead telluride segmenting for improved thermoelectric efficiency', *Proceedings of IEEE/AIAA Thermoelectric Specialists Conference*, (1966), paper 12, pp. 1–9.
4. P. Rouklove and V. Truscello, 'Performance testing of thermoelectric generators at JPL', *Proceedings of 10th IECEC*, (August 1975), Newark, Delaware, pp. 873–9.
5. J.C. Bass and N.B. Elsner, 'Segmented selenide thermoelectric generator', *Proceedings of Third International Conference on Thermoelectric Energy Conversion*, (March 1980), University of Texas at Arlington, pp. 8–12.
6. P. Rouklove, 'Design and analysis of a cascaded thermoelectric generator', *Proceedings of 6th IECEC*, (August 1971), Boston, Massachusetts, pp. 611–20.
7. H.J. Goldsmid, *Applications of Thermoelectricity*, (Ed.) B.L. Worsnop, Methuen monograph, London, 1960.
8. V.S. Ramanychev, 'Volatility of semiconductor materials in a gaseous medium (T/E)', *Toplofizika Vysokikh Temperatur*, **1** (3)(1963), 368–75.
9. D.R. Field and G.W. Bunde, 'A model to describe long-term performance characteristics of pressure contacted lead-telluride thermoelectric generators', *Proceedings of 4th IECEC*, (September 1969), Washington DC, pp. 425–35.
10. J.W. Killian, 'A method to arrest the weight loss of PbTe', *Proceedings of 3rd IECEC*, (August 1968), Boulder, Colorado, pp. 272–6.
11. E.A. Skrabek, 'Improved long term performance of TAGS thermoelements', *Proceedings of 9th IECEC*, (August 1974), San Francisco, California, pp. 160–4.
12. H.G. Stanley, L.H. Rovner, D. Snowden and N.B. Elsner, 'Vaporization and compatibility of SiGe radioisotope thermoelectric generator', *Proceedings of 7th IECEC*, (September 1972), San Diego, California, pp. 159–67.
13. N.B. Elsner, 'Ion plating of Si–Ge thermoelectric couples with Si_3N_4 for vaporization suppression and electrical insulation', *Proceedings of 9th IECEC*, (August 1974), San Francisco, California, pp. 152–5.
14. E. de Winter, 'Use of protective xenon atmospheres in silicon germanium thermoelectric generators', *JPL Internal Engineering memo*, (March 1972), 342–169 (900–546).
15. W.J. Bifano, 'Hybrid TE panel test results', *Proceedings of 7th IECEC*, (September 1972), San Diego, California, pp. 194–9.
16. N.I. Erzin and N.V. Makov, 'Bonding of thermoelements in thermoelectric devices', English translation, *Applied Solar Energy (USA)*, **1**(5), (1965), 32–9.
17. D. Ilzycer, A. Sher and M. Shiloh, 'Electrical contacts to thermoelectric bismuth telluride based alloys', *Proceedings of Third International Conference on Thermoelectric Energy Conversion*, (March 1980), University of Texas at Arlington, pp. 200–2.
18. M. Lanxner, I. Shai, A. Brandstetter and G. Vekutieli, 'High reliability contacts for miniature thermoelectric converters', *Proceedings of 13th IECEC*, (August 1978), San Diego, California, pp. 1984–8.
19. G. Stapfer and P. Rouklove, 'The thermoelectric generator test program at JPL', *Proceedings of 7th IECEC*, (September 1972), San Diego, California, pp. 149–58.
20. U. Birkholz, 'High temperature thermocouples from GeSi mixed crystals for thermoelectric converters', *Zeitschrift für angewandte Physik*, **22**, pp. 395–8.
21. J.D. Gylfe and J.H. Van Osdal, 'Status report on small reactor thermoelectric power systems for unmanned space applications', *Proceedings of 4th IECEC*, (September 1969), Washington, DC, pp. 300–7.
22. V. Raag and R.F. Berlin, 'A silicon–germanium solar thermoelectric generator', *Energy Conversion*, **8**(1968), 161–8.

23. D. Wright, 'Materials for direct conversion thermoelectric generators', *Metallurgical Reviews*, **15** (1970), 147–60.
24. J.P. Dismukes et al., 'Thermal and electrical properties of heavily doped GeSi alloys up to 1300 K', *Journal of Applied Physics*, **10** (1964), 2899–907.
25. M. Piper, 'Germanium–silicon thermoelectric material development', *Proceedings of UKAEA–ENEA, International Symposium on Industrial Applications for Isotopic Power Generators*, (1966), UKAEA, Harwell, pp. 349–59.
26. N.B. Elsner, G.B. Bradshaw, C.H. Shearer and C.C. Morris, 'Fabrication and testing of a thermoelectric ring converter segment for space applications', *Proceedings of 9th IECEC*, (August 1974), San Francisco, California, pp. 165–70.

7

Thermoelectric Refrigerators

7.1 INTRODUCTION

WHEN a d.c. current is passed through a thermocouple it behaves as a heat pump, and if used to pump heat from a low temperature junction to the high temperature junction the device acts as a refrigerator. However, its coefficient of performance is low compared to that of conventional compression-type refrigerators, as shown in Figure 7.1,[1] and consequently thermoelectric refrigerators are usually employed in applications where properties such as compactness and ability to function in vacuum or the absence of harmful gases are considered of overriding importance. Another important factor in their favour is that the coefficient of performance is independent of the size of the system to be cooled, a property which particularly favours miniature low capacity cooling devices.

7.2 THERMOELECTRIC MODULE

The main component of a thermoelectric refrigerator is the module. This is identical in construction to a thermoelectric generator module. All the cold junctions are on one face and the hot junctions on the other. Heat is pumped to the hot junction at a rate which is proportional to the current through the circuit and to the number of couples. Materials most widely used in thermocouple construction are bismuth telluride or an alloy of bismuth, tellurium, selenium and antimony. The relatively low melting point of these materials restricts the hot side temperature of operation to about 400 K, although this is ample for most refrigeration applications. High temperature thermoelectric cooling modules have been developed[2] which can operate at temperatures as high as 550 K, which is outside the range of compression refrigerant systems. Modules are available off the shelf in a variety of sizes and heat pumping capacities and in single or multiple arrangements as shown in Figure 7.2. Single layer sizes range from 2.5 mm to 50 mm square and from 2.5 mm to 7.5 mm thick, with cooling capacities ranging from a few watts to >130 watts. Multistage modules of up to six layers can generate several hundred watts of cooling capability and provide temperatures as low as 173 K. Because the lower stages pump heat dissipated by the stages above them as well as from the load, multistage modules take on a typically pyramidal shape. The choice of module depends upon the intended application, and compre-

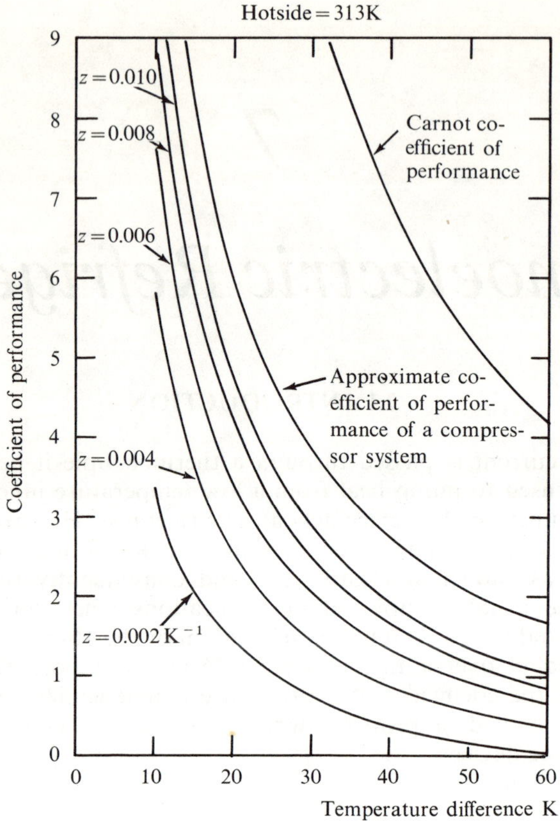

Figure 7.1 Maximum coefficient of performance versus temperature drop for a Peltier couple. (After Cadoff and Miller, reference 1.)

hensive data/information sheets are available from manufacturers to assist customers in making a choice.† Both the cost and coefficient of performance of a module increase with the number of stages, consequently in commercial applications there will be a trade-off between the two. The cold side temperature as a function of number of stages for a 'MELCOR' refrigerator module is shown in Figure 7.3. It is apparent that a greater temperature differential can be reached by operating the device in vacuum because the heat load is less in vacuum than in dry air or nitrogen.

7.3 POWER SUPPLY

Peltier devices operate at high currents (typically 1–15 amps) and low voltages (from a fraction of a volt upwards), and clearly a thermoelectric generator

† Borg–Warner Thermoelectrics, 3570, North Avondale Avenue, Chicago, IL 60018.
Cambridge Thermionic Corp. (Cambion), 445 Concord Ave, Cambridge MA 02238.
Marlow Industries Inc., 1021 South Jupiter Road, Garland, TX 75042
Materials Electronic Products Corp. (MELCOR), 990 Spruce Street, Trenton, NJ 08648.
Mectron (Frigistor) Ltd, 480 Bath Road, Slough, Bucks, SL1 0BL.

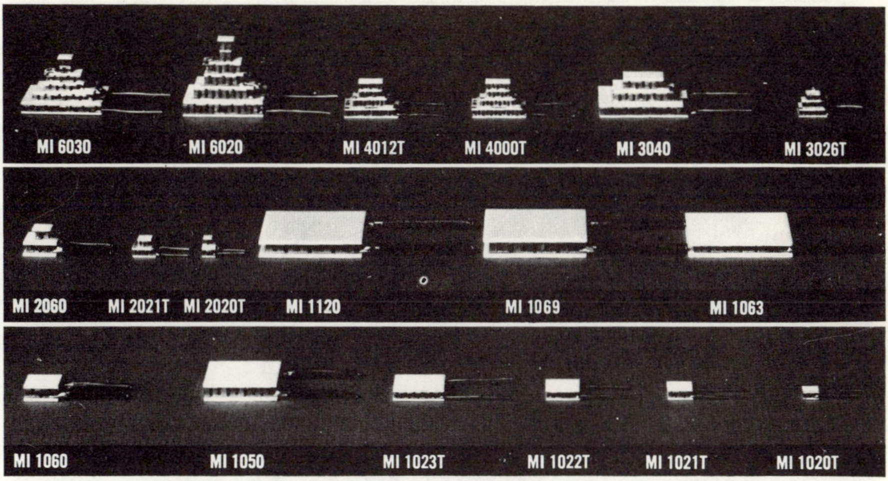

STANDARD THERMOELECTRIC HEAT PUMPS

MODEL	I_{max} (amps)	V_c @ I_{max} (volts)	TOP CERAMIC L (inches)	TOP CERAMIC W (inches)	BOTTOM CERAMIC L (inches)	BOTTOM CERAMIC W (inches)	HEIGHT H (inches)	$T_{c\ max}$ VACUUM (10^{-6} Torr)	$T_{c\ max}$ DRY NITROGEN (14.7 psia)	HEAT PUMP CAPACITY (mw/°C)
MI 1020T	2.0	0.8	.156	.156	.156	.156	.085	−41	−34	17
MI 1021T	2.0	2.0	.260	.260	.260	.260	.085	−41	−34	39
MI 1022T	2.0	3.6	.346	.346	.346	.346	.085	−41	−34	73
MI 1023T	2.0	8.2	.518	.518	.518	.518	.085	−41	−34	164
MI 1050	5.0	3.7	.750	.750	.750	.750	.178	−41	−34	218
MI 1060	6.0	0.7	.380	.380	.380	.380	.172	−41	−34	46
MI 1063	6.0	7.5	1.167	1.167	1.167	1.167	.172	−41	−34	470
MI 1069	6.0	15.0	1.500	1.500	1.500	1.500	.172	−41	−34	841
MI 1120	12.0	3.4	1.167	1.167	1.167	1.167	.195	−41	−34	450
MI 2020T	1.7	0.9	.100	.100	.156	.156	.149	−70	−56	4
MI 2021T	1.5	2.0	.156	.156	.260	.260	.149	−64	−56	11
MI 2060	4.7	0.8	.207	.207	.380	.380	.304	−70	−56	10
MI 3026T	1.2	2.9	.174	.160	.346	.346	.234	−84	−62	6
MI 3040	3.5	6.5	.511	.339	1.113	.855	.436	−80	−68	46
MI 4000T	0.5	6.0	.314	.160	.678	.522	.345	−85	−50	4
MI 4012T	1.0	6.0	.314	.160	.678	.522	.345	−88	−60	8
MI 6020	1.7	6.0	.200	.200	1.113	.855	1.012	−100	−33	3
MI 6030	3.2	6.0	.200	.200	1.113	.855	.832	−105	−56	4

Courtesy Marlow Industries, Garland, Texas.

Figure 7.2 Standard thermoelectric heat pumps. (Courtesy Marlow Industries).

would provide an ideal supply for a thermoelectric refrigerator.[3] The input to the refrigerator should be direct current (d.c.). An alternating component to the current would be undesirable on two accounts; firstly it would produce Joule heat, and secondly it would not pump heat. The maximum cooling that can be achieved by Peltier cooling can be increased momentarily if the device, when operating in its cooled down steady state condition, is subjected to high current pulses of short duration. Any joule or resistance heating developed in the device requires a finite time to be conducted to the cold junction, whereas any additional Peltier cooling is generated instantly. This technique was first reported more than 20 years ago by Stil'bans[4] and subsequently investigated exhaustively.[5–13] Pulsed cooling offers advantages in some specific applications such as infra-red detectors. Novel geometries for the thermoelement legs have also been proposed and investigated, the objective being to overcome the inherent limitations of devices with uniform geometry.[14–16]

The current supply usually used in practice is rectified alternating current.

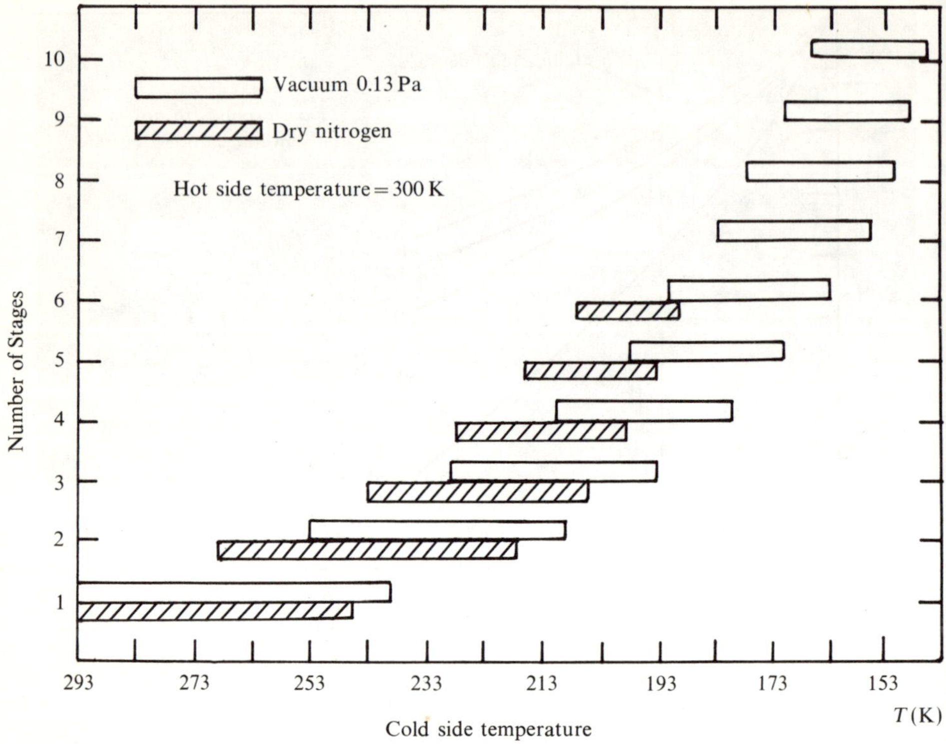

Figure 7.3 Cold side temperature as a function of number of stages, (EDN 20 May 1980).

Full wave rectification can be achieved using a simple circuit consisting of a transformer and two silicon diodes. The amount of filtering used then depends upon the particular application, with a 10 per cent ripple resulting in a loss in efficiency of about 4 per cent. The loss in performance due to imperfect filtering is taken into account through a so-called form factor F where $F = I_{\text{r.m.s}}/I_{\text{d.c.}}$ and $I_{\text{r.m.s.}}$ in the root-mean-square value of the current. The effect of a form factor differing from unity is to decrease the Seebeck coefficient from α to α/F, with consequent effects on the maximum coefficient of performance as indicated in Figure 7.4.[17] Also shown is the F for an unfiltered full wave, and it is apparent that for small temperature differences even a small amount of filtering will give a reasonable approach to d.c. performance. Commercial power supplies can be used; those with a variable output can be used with a wide range of thermoelectric devices. Fixed output supplies are also available for use with specific devices, and if required the d.c. output voltage can be varied using an autotransformer.

The highest current applied to the module under ideal conditions to obtain the lowest cold junction temperature is called the optimum current. Modules are manufactured with different optimum ratings, although in practice they are used below this current value. The price per watt of heat pumped is less for high current rating modules, and as low current rating modules have the highest heat flux, these should be employed with small cooling loads while those with high current ratings should be used for larger cooling loads.

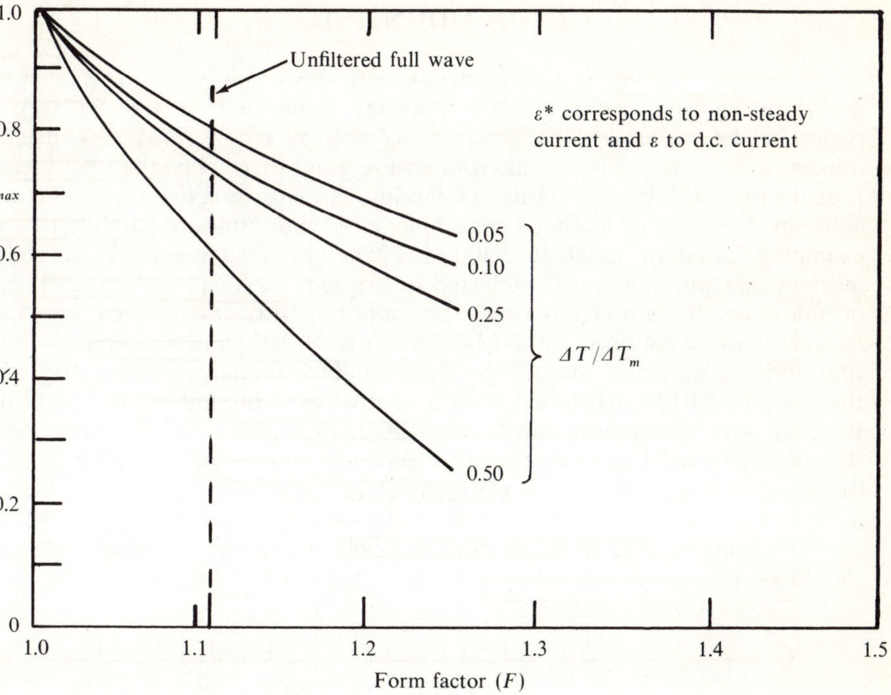

Figure 7.4 Form factor dependence (for different $\Delta T/\Delta T_m$) of maximum coefficients of performance. (After Cadoff and Miller, reference 1.)

7.4 HEAT SINK

The heat which is being removed from the cold environment is deposited on the hot side of the module and must subsequently be dissipated into the surroundings. Evidently the heat sink must be at a higher temperature than ambient, and a difference of 5 to 15 K is found practical. Heat sinks fall into three main categories—natural convection, forced convection, and liquid cooled—and their performance is measured in kelvin per watt. Choice of sink depends primarily upon the heat load to be dissipated, with typical performance or conductance values of 2.0 to 0.5 K/W, 0.5 to 0.02 K/W and 0.02 to 0.005 KW for natural convection, forced convection and liquid cooling, respectively. Natural convection has the advantage of no moving parts, and although it can be employed with large area heat sinks when the cooling load is moderately low (less than 50 watts) it is generally inadequate for most thermoelectric applications. Since the performance of a cooling module depends on the rate of heat dissipation from the heat sink, the potential of a device is best realized when used in conjunction with a liquid-cooled heat sink. Forced air cooling, although not as effective as liquid cooling, is an alternative which can be employed in situations where for some reason or other, cooling with a liquid is not possible. Vortex cooling tubes, which use adiabatically expanded air rather than water as a coolant, have also been used.[18]

7.5 MOUNTING

A thermoelectric device is only as strong as the semiconductor thermoelements and should not be used as a load supporting structure. Heat transfer from the region to be cooled to the ambient, as well as to the heat sink itself, is important and care must be taken to ensure good thermal contact between all heat transfer surfaces, including the module faces (in practice very difficult to achieve). A variety of methods are employed in mounting the module (see, for example, Cambion catalogue 300, p. 25). The recommended method is mechanical clamping; this is preferred to methods such as bonding with resins or soldering. If clamping is used, the mounting surfaces between which the module is to be clamped should be ground to ± 0.025 mm and coated with a thin film of thermal contact grease. Matched modules having the same thickness (± 0.013 mm) should always be used in multimodule configurations. Bonding with epoxy resin can be used in small heat load applications, but it should not be used in vacuum and is generally not recommended because of the rigid nature of the bond. Component temperatures of a normal cooling module should not exceed 400 K. Although low melting point solders (<400 K) can be used in mounting thermoelectric devices, soldering methods should be employed only where alternatives are not practical.

7.6 APPLICATIONS

Because the coefficient of performance of a thermoelectric refrigerator is independent of its size, devices based upon the Peltier effect are particularly attractive in miniature applications using single thermocouples or modules consisting of a small number of thermocouples. A number of the first applications of thermoelectric cooling have been described in standard texts[19,20] and range from dew point hygrometry[21,22] to thermoelectric thermostats.[23,24] Applications have also been found in cooling expansion cloud chambers,[25] in providing cold traps for vacuum[26] systems, and in zone melting.[27] Thermoelectric cold stages for microtomes and microscopes possess the additional benefit of serving as hot stages when the current direction is reversed. This facility has been further developed in applications such as hot and cold plates and hot and cold liquid baths which provide fixed temperatures in the range $-10°C$ to $+50°C$.[28] Thermoelectric immersion coolers have also been described. A typical cooler (manufactured by Mectron) extracts 52 watts (175 Btu per hour) at a power input of 70 watts and can establish a 50 K temperature difference between a thermally insulated bath and heat sink which has circulating water. The possibility of using thermoelectric refrigeration at very low temperatures has also been considered, and it is thought that semiconductor thermoelements could produce temperatures substantially lower than those reached by a helium dilution refrigerator.[29]

Thermoelectric coolers have also found application in maintaining temperature-controlled chambers, domestic refrigerators, air cooling panels[2] and in solar-assisted air conditioners[30] and refrigeration for medical stores.[31] Solid state air conditioners have been developed with multi-hundred watts cooling capacity.[32] The Borg–Warner AHP-1700 is shown in Figure 7.5 and

Figure 7.5 Solid state air conditioner AHP 1700. (Courtesy Borg–Warner Thermoelectrics.)

its performance characteristics in Figure 7.6. The nominal heat pumping capacity at zero ΔT is 320 W in a room temperature ambient of 25°C. The thermal load capacity, when providing 20 K of cooling, is 160 W at a 25°C ambient. Suitable applications are in cooling computers and machine tools. A large scale gas-to-gas cooling system has also been described which is intended for air conditioning on electric passenger trains.[33,34]

In recent years the cost of cooling modules has dropped dramatically and new applications have been found in cooling electronic components to control

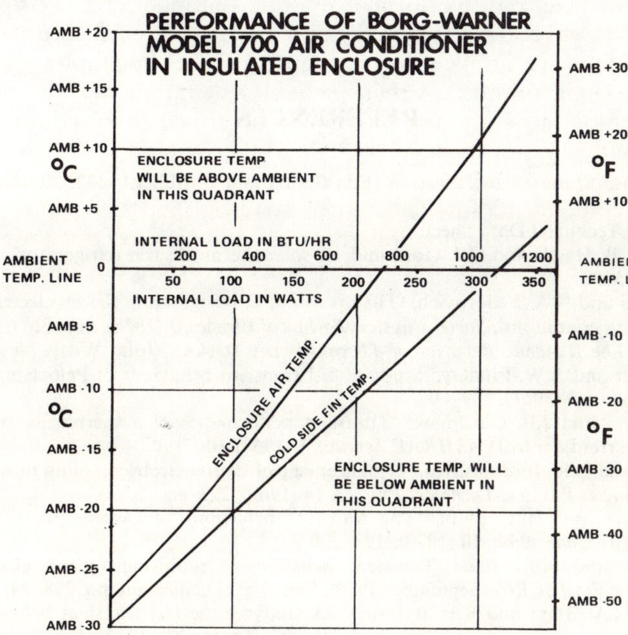

Figure 7.6 Performance characteristics of the AHP-1700. (Courtesy Borg–Warner Thermoelectrics.)

critical parameters such as noise or bias current, or providing safe operating temperatures for semiconductor devices in ambient temperatures above the safe limits of the device.[35,36] Miniature thermoelectric coolers have been used to cool microchips.[35] As the dimensions of modern electronic devices decrease and the systems become more compact, the dissipation of internally generated heat becomes an increasing problem. Thermoelectric devices (TEDs) fabricated by vacuum deposition techniques have been successfully employed in reducing the temperatures of miniature electronic devices such as charge-coupled devices (CCDs) to more than 30 K below ambient.[37] A reduction in temperature considerably improves the performance of electronic devices[38] and in memory applications leads to an increase in the number of line elements and associated storage capacity. Thermoelectric cooling is also finding increased application in electro-optical systems such as infra-red detectors. In detectors requiring an operating temperature of down to 193 K the device can be cooled thermoelectrically rather than with liquid air.[39] Thermoelectric coolers are also employed in providing black body reference source for infra-red detecting systems used in thermal mapping.[40]

Another application is in the construction of differential temperature sources for testing thermal imagers. The simplest form of this device consists of a 'sandwich' in which the thermoelectric elements form the filling between two copper plates. The front plate has a pattern of holes or slots cut in it. If by using the thermoelectric elements a small temperature difference is set up between the two plates, the performance of a thermal imager can be measured by noting how small a temperature difference it can detect. This depends upon the size of the whole pattern, so that one can obtain the relation between the temperature and spatial resolution which determines the quality of the thermal image.[41]

REFERENCES

1. *Thermoelectric Materials and Devices*, (Ed.) Cadoff and E. Miller, p. 322, Reinhold, London, 1960.
2. 'MELCOR', Technical Data Sheet.
3. G.J. Vella, L.B. Harris and H.J. Goldsmid, 'A solar thermoelectric refrigerator', *Solar Energy*, **18** (1976), 355–9.
4. L.S. Stil'bans and N.A. Fedorovich, 'The operation of refrigerating thermoelectric elements in non-stationary conditions', *Soviet Physics, Technical Physics*, **3** (1958), 460–3.
5. P.E. Gray, '*The Dynamic Behavior of Thermoelectric Devices*', John Wiley, New York, 1960.
6. K. Landecker and A.W. Findlay, 'Study of fast transient behaviour of Peltier junctions', *Solid State Electronics*, **3** (1961), 239–60.
7. W.F. Stoecker and J.B. Chaddock, 'Transient performance of a thermoelectric refrigerator under step-current control', *ASHRAE Journal*, **5** (1963), 61–7.
8. V.P. Babin and E.K. Iordanishvili, 'Enhancement of thermoelectric cooling in non-stationary operation', *Soviet Physics, Technical Physics*, **14** (1969), 293–8.
9. R.P. Bywaters and H.A. Blum, 'The transient behaviour of cascade thermoelectric heat pumps', *Energy Conversion*, **10** (1970), 193–200.
10. G.E. Hoyos and K.R. Rao, 'Transient behavior of multistage thermoelectric coolers', *Proceedings of 7th IECEC*, (September 1972), San Diego, California, pp. 228–34.
11. R.L. Field, H.A. Blum and R.P. Bywaters, 'A study of the fast transient behavior of pulsed thermoelectric coolers', *Proceedings of 7th IECEC*, (September 1972), San Diego, California, pp. 235–40.
12. G.E. Hoyos, K.R. Rao and D. Jerger, 'Numerical analysis of transient behavior of thermoelectric

coolers', *Proceedings of First International Conference on Thermoelectric Energy Conversion*, University of Texas at Arlington (September 1976), 76–89.
13. G.E. Hoyos, '*Fast Transient Response of Novel Peltier Junctions*', PhD Dissertation, Electrical Engineering Department, University of Texas at Arlington, Texas, 1976.
14. P.W. Cowling, J.E. Sunderland and L.J. Ybarrondo, 'Development of a modified cascade thermoelectric heat pump', *Energy Conversion*, **9** (1969), 107–17.
15. K. Landecker, 'Improvement of the performance of Peltier junctions for thermoelectric cooling', *Journal of Physics: C, Solid State Physics*, **3** (1970), 2146–50.
16. K. Landecker, 'Some further remarks on the improvement of Peltier junctions for thermoelectric cooling', *Energy Conversion*, **14** (1974), 21–33.
17. R.H. Vought, 'Design Calculations for Peltier Cooling'. In, *Thermoelectric Materials and Devices*, (Ed.) Cadoff and Miller, Chapter 17, Reinhold, London, 1960.
18. K. Landecker, 'The applications of the "vortex" cooling tube to the cooling of thermojunctions', *Proceedings of Second International Conference on Thermoelectric Energy Conversion*, (March 1978), University of Texas at Arlington, 134–6.
19. A.F. Ioffe, '*Semiconductor Thermoelements and Thermoelectric Cooling*', Infosearch, London, 1957.
20. H.J. Goldsmid. In '*Applications of Thermoelectricity*', (Ed.) B.L. Worsnop, Methuen Monograph, London, 1960.
21. H.L. Penman, '*Humidity*', Institute of Physics, London, 28, 1955.
22. D.A. Wright, *Semiconductors*, 3rd Edition, Methuen, London, 49, 1958.
23. R.A. Valitov and A.I. Aleksandrov, 'Termostat S primeneniem poluprovodnukov', *Izmeritel'naya tekhnika*, pp. 64–5, 1957.
24. E.K. Iordanishvili and L.G. Tkalich, 'Poluprovodnukovsky termostat dlya avtogeneratorov', *Zhur Tekh*, **27** (1957), 1215–20.
25. D. White et al., 'A thermoelectric cooled expansion cloud chamber', *Proceedings of Second International Conference on Thermoelectric Energy Conversion*, (March 1978), University of Texas at Arlington, pp. 41–2.
26. E.A. Kolenko, *Vestnik Akademii nauk SSSR*, **27** (1957), 50.
27. W.G. Pfann, K.F. Benson and J.H. Wernick, 'Some aspects of Peltier heating at liquid–solid interfaces in germanium', *Journal of Electronics*, **2**(6) (1957), 597–608.
28. Cambion, Catalog 300 data sheet.
29. H.J. Goldsmid and A.S. Gray, 'Thermoelectric refrigeration at very low temperatures', *Cryogenics*, (1979), 289–92.
30. R.N. Basu, 'A solar-assisted thermoelectric air-conditioner, humidifier, (SATEACH)', *Proceedings of Third International Conference on Thermoelectric Energy Conversion*, (March 1980), University of Texas at Arlington, pp. 117–9.
31. R.L. Field, 'Photovoltaic/thermoelectric refrigerator for medicine storage for developing countries', *Solar Energy*, **25** (1980), 445–7.
32. Borg–Warner, Thermoelectric Data Sheet.
33. J.G. Stockholm and J. Buffet, 'Large gas to gas thermoelectric heat pumps', *Proceedings of Third International Conference on Thermoelectric Energy Conversion*, (March 1980), University of Texas at Arlington, pp. 120–6.
34. J.G. Stockholm 'Prototype air conditioning unit for railway passenger coaches'. *Proceedings of Fourth International Conference on Thermoelectric Energy Conversion*, (1982), University of Texas at Arlington, 136–41.
35. D.A. Zeskind, 'Thermoelectric heat pumps cool packages electronically', *Electronics*, **31** (1980), 109–13.
36. J. McDermott, 'Thermoelectric coolers tackle jobs heat sinks can't', EDN, May 20, (1980), 111–7.
37. J.W. Harpster, 'Vacuum deposited TED's for electronic devices chip cooling', *Proceedings of Second International Conference on Thermoelectric Energy Conversion*, (March 1978), University of Texas at Arlington, pp. 43–7.
38. J.W. Harpster, et al., 'Integrated thermoelectric devices; manufacturing methods study', *Final Report, Contract N00019–76–C–0258*, (January 1977).
39. 'A Thermoelectric Refrigerator', *CVD News*, No. 23 (November 1980).
40. P.B. Click, 'Application of thermoelectric coolers as blackbody reference sources', *Proceedings of First International Conference on Thermoelectric Energy Conversion*, (September 1976), University of Texas at Arlington, pp. 168–74.
41. E.H. Putley, RSRE, Malvern, Private communication.

8

Thermoelectric Generators

8.1 INTRODUCTION

PROVIDED that a temperature difference can be maintained across the thermoelements of a generating device, it will deliver power to an external load. Generators can be classified into two types: isotopic and non-isotopic. Although in principle non-isotopic generators are relatively straightforward and unsophisticated, with heat at the hot junction being provided by a variety of sources (fossil flame, solar radiation, oceanic temperature gradients and so on), in practice this is not so. For example, the effort needed to develop multifuelled burners greatly exceeded that required for the whole of the rest of the system. Where temperature differences are small, there is a problem in avoiding excessive temperature drop in the electrical insulation that must be interposed between the heat source and the hot face of the module and between the cold face of the module and the heat sink. The concept of employing isotopic energy as a heat source has resulted in a proliferation in thermoelectric generator applications. It is clear from Figure 8.1[1] that the long lives of isotopes, coupled with their high energy density, facilitate the production of generators that can operate unattended over long periods.

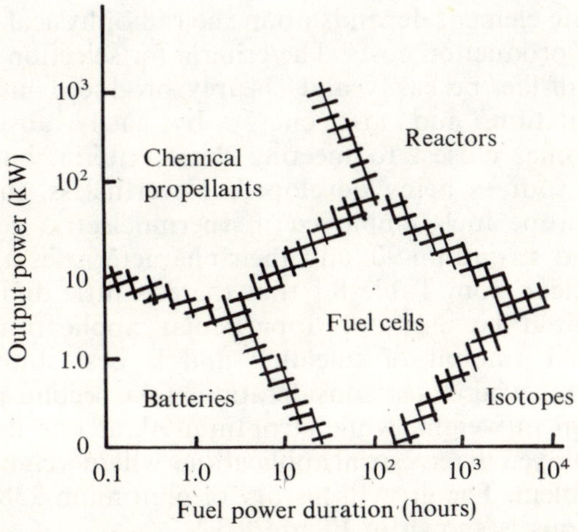

Figure 8.1 Areas of superiority of energy sources. (After Penn, reference 1.)

However, unlike fossil fuel heat sources, the rate of isotopic energy release is not controllable. The isotopes disintegrate or decay spontaneously and power is delivered continuously via the nuclear radiations, α, β, γ rays and neutrons. Although there is no way of knowing which nuclei in the source decay at any instant in time, the number N_t remaining after a time t is given by $N_t = N_0 \exp(-\lambda t)$, where N_0 is the number present at time $t=0$ and λ is the decay constant. The power of the source thus decreases exponentially with time. A parameter used as a measure of the 'life-time' of a radioactive source is the half-life $t_{\frac{1}{2}}$. This is defined as the time taken for half the original number of nuclei present to decay.

All radiations are potentially able to damage biological tissue; consequently, all sources must be shielded. The radiations, although penetrating, are emitted at relatively low power density, whereas thermoelements are essentially high-thermal-flux devices. In low-power systems, such as the 1 W generator shown in Figure 8.2,[2] the thermoelements cannot cover all the available surface of the source, and thus good thermal insulating materials capable of operating at high temperatures are required. The larger the radioisotope heat source, the higher the surface power density; consequently, in higher-powered systems, such as the UKAEA's RIPPLE 30 W generator, most of the surface of the source capsule is in contact with the thermoelectric modules, and thermal insulation plays only a minor role. As indicated previously, thermoelectric power is generated at relatively high values of current and low voltages, while the majority of devices require inputs of several volts. Consequently, d.c.–d.c. converters are usually employed, which brings the number of essential components of an isotope powered generator to five: heat source, shielding and containment, thermal insulator, thermoelectric module and d.c.–d.c. converter.

8.2 ISOTOPE HEAT SOURCE

Of the 1000 or so radioactive isotopes that have been identified to date, about 50 possess half-lives in the range of 100 days to 100 years, and of these about a dozen can be produced in reasonable quantities. Selection of a suitable isotope for a thermoelectric element depends upon the radiophysical properties as well as the source and production costs. The criteria for selection are that it should possess a long half-life, be easily and cheaply produced in stable form, lend itself to encapsulation, and lose energy by easily absorbed radiations. Plutonium-238 comes closest to meeting these criteria, but its expense has resulted in other sources being developed. Nevertheless, there are only two principal radioisotope fuels employed in thermoelectric generators, namely plutonium-238 and strontium-90, and their characteristics are summarized in Table 8.1.[3] It is clear from Table 8.1 that there is little difficulty in choosing which source should be used for a particular application. Plutonium-238 requires a minimal amount of shielding and is best suited for space and medical applications where cost considerations take second place to compactness and reduction in weight, while strontium-90, at one fortieth the cost of plutonium, is employed in terrestrial applications where weight of shielding does not present a problem. The growth history of plutonium-238 as a nuclear fuel in space applications is shown in Figure 8.3.[4]

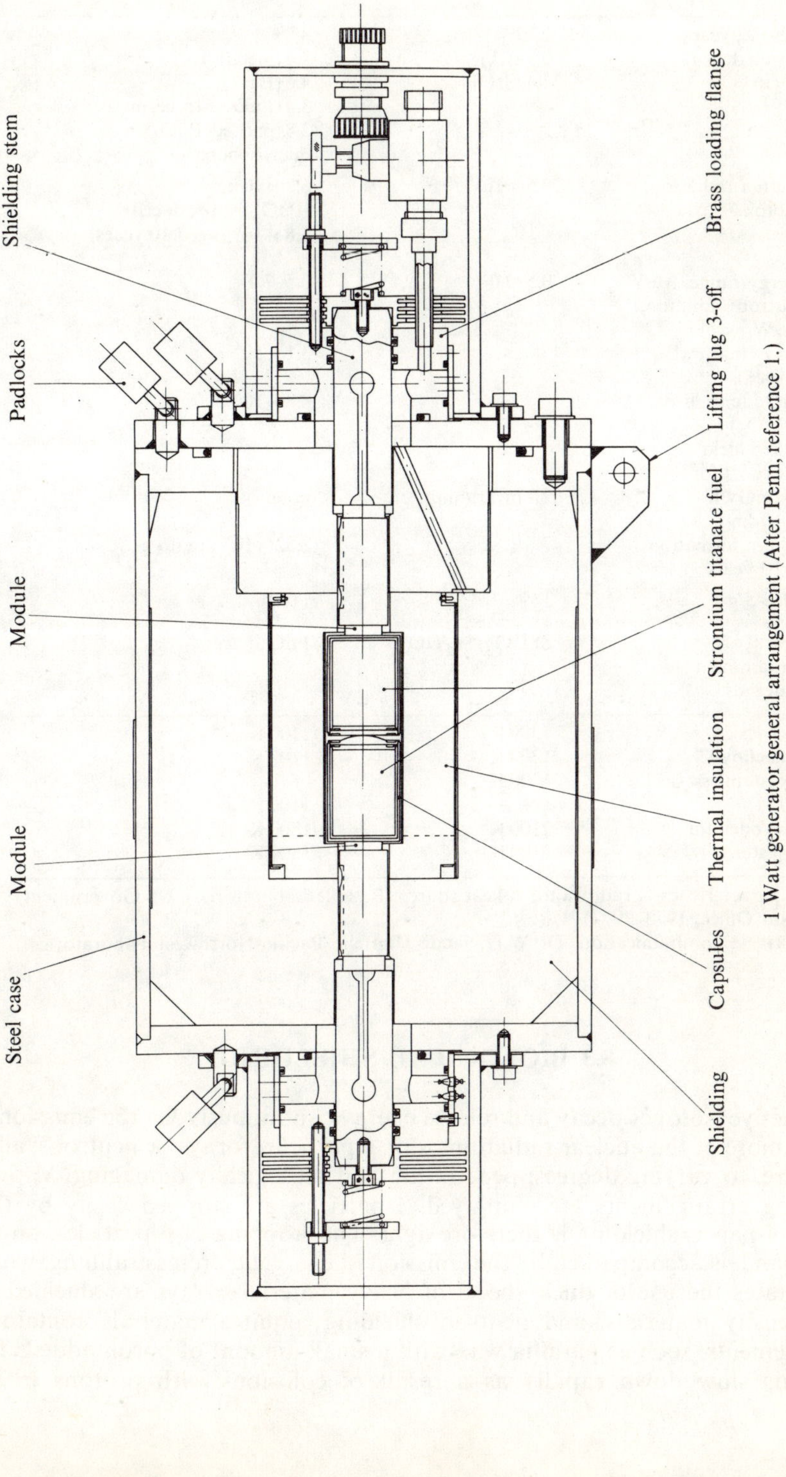

Figure 8.2 One watt thermoelectric generator, general arrangement.

Table 8.1 *Heat source characteristics* (After Carney, ref. 3)

	^{90}Sr	80% ^{238}Pu
half-life, years	27.7	87.8
power density, W/cm³	0.9–1.1 (SrO-TiO_2) 1.4 (SrO)	7.4 (metal) 4.6 (PuO_2) 3.4 (PuO_2-Mo cermet) 2.8 (packed PuO_2) microspheres
gamma radiation, photons/s-W	9.5×10^{10}	6.3×10^5 (PuO_2 as produced) 1.8×10^6 (aged 10 years)
energy range, MeV	0.5–2.0	0.5–7.0
neutron radiation, n/s-W		
PuO_2 (natural oxygen)	—	3.1×10^4
PuO_2 (enriched in $^{16}O_2$)	—	8.8×10^3
10 W shield thickness for 200 mGy/h at 10 cm	2.1 in uranium	none
helium formation in 10 years	—	2.52×10^{-1} mol/kg
price \$/g	14†	1100*
typical containment materials	$SrTiO_3/Sr_2TiO_4$	PuO_2
Haynes-25	1300 K	1100 K
Hastelloy-C	1300 K	1100 K
316 stainless steel	1300 K	—
T-111	—	1100 K
molybdenum	2100 K	1800 K
tungsten	2100 K	1800 K

†Characteristics of radioisotopic heat source (Battelle Laboratories), US Government Print Office, 1973–797–171/61
*Private communication, Dr W.E. Saude (Battelle Pacific Northwest Laboratories)

8.3 BIOLOGICAL SHIELDING

Radioactive isotopes decay and release energy continuously via the emission of one or more of the nuclear radiations α or β particles, γ-rays or neutrons; all of these are, to varying degrees, penetrating and biologically damaging. Various shielding arrangements are employed. α-particles are stopped easily by thin sheets of paper; shielding is therefore light. The stopping of β-particles, on the other hand, is accompanied by the emission of energetic bremsstrahlung, which necessitates the use of thick sheets of heavy material. γ-rays are shielded by high-density materials, and neutron shielding requires materials containing light elements, such as paraffin wax with a small amount of boron added. The neutrons slow down rapidly as a result of collisions with protons in the

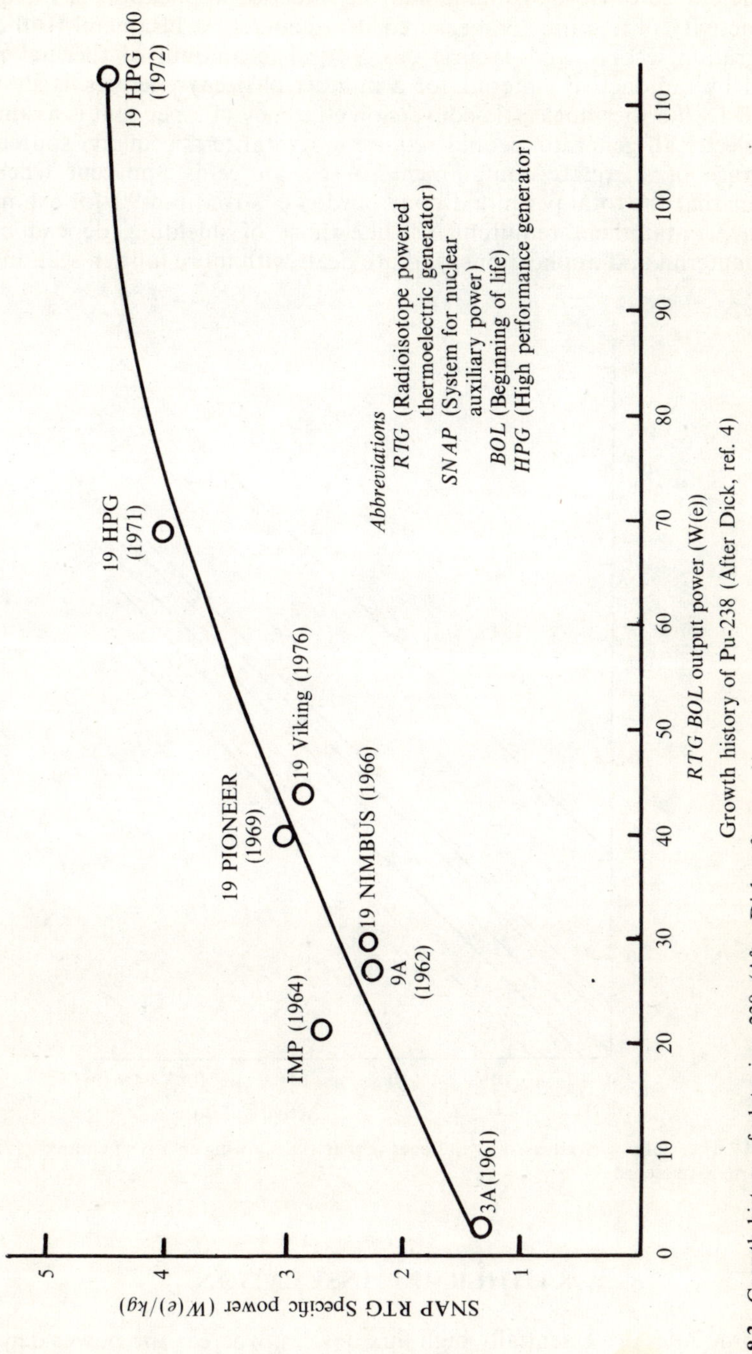

Figure 8.3 Growth history of plutonium-238. (After Dick, reference 4.)

paraffin and are then absorbed by the boron.[10] The shielding thus depends upon the choice of isotope source and the intended application of the device.

The activity of a source is measured in becquerels; a becquerel (Bq) is one disintegration/second, and 1 curie ≡ 37 GBq. The amount of thermal power released by radioactive material, for a number of decay energies, is shown in Figure 8.4.[3] If a thermoelectric-conversion efficiency of 5 per cent is assumed, a 10 W (electrical) generator would require a several terabecquerel source. The importance of adequate containment of the source is apparent when one considers that the total permitted body burden of strontium-90, for example, is 750 kBq. Containment requirements, like those of shielding, depend on the source material and applications, and are dealt with more fully in section 9.3.1.

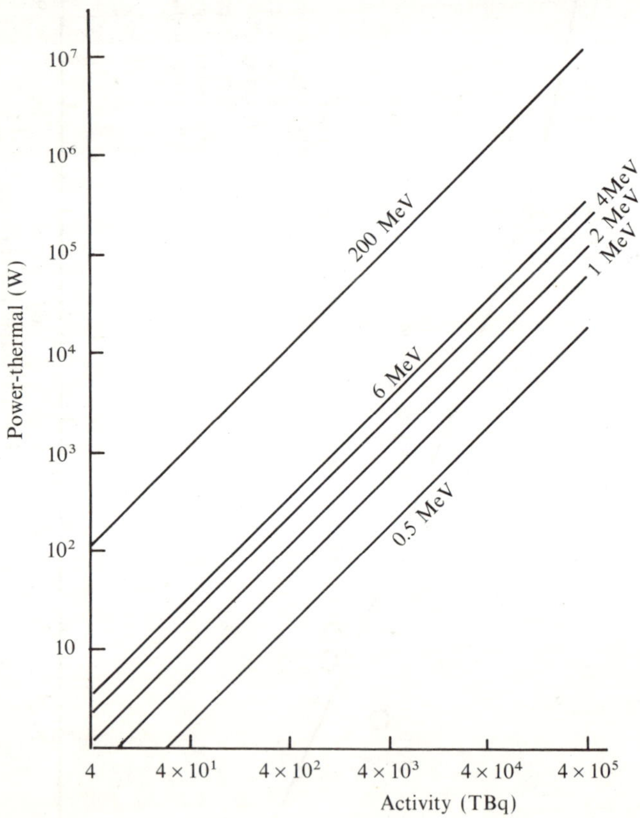

Figure 8.4 Thermal power released by radioactive material showing energy of emitted radiations. (After Carney, reference 3.)

8.4 THERMAL INSULATION

Thermomodules are essentially high-flux devices, whereas the power density of isotope sources is relatively low. To insulate source areas not in contact with thermoelements, special high-temperature insulating materials have been developed. The total heat conduction through a material is the summation of several heat transfer processes, i.e., radiation transmission, radiation conduction, solid conduction, gas convection, with the dominant mechanism depending on the temperature.

In space applications, heat losses are mainly radiative, and fibrous and multifoil insulations have been developed. The multifoil system consists of many layers of thin metallic foils, which serve as radiative barriers, and which are separated from one another by high-purity refractory oxide particles.[5] Fibrous insulations, such as MIN-K, are fabricated from submicron silica particles, opacifiers and fibrous reinforcement. A number of formulations are employed, depending on strength requirements and the temperature of operation.[6] The MIN-K insulation system has been recommended for use in the development of low cost high efficiency generators.[7]

In the UK a very low thermal conductivity material (microtherm) has been developed at Harwell. Initially tailored to suit a RIPPLE generator hot-face temperature of 600 to 900 K, it now covers temperatures up to 1400 K.[8] Very similar to MIN-K, the material possesses a pore size smaller than the mean free path of the gas molecules; thus cavity-wall scattering will give rise to a thermal conductivity value that is less than that of still air. Radiation-transmission inhibitors (opacifiers) are also introduced, and the powder is bound with an inorganic fibre.

8.5 THERMOELECTRIC MODULE

The arrangement of thermocouples within a module has been discussed in section 6.3. The thermoelectric efficiencies of modules fabricated from established materials are shown in Figure 8.5. A 10 per cent electrical contact loss and a thermal insulation bypass loss corresponding to 20 per cent of the thermoelectric module cross section is assumed.

A typical example of a thermoelectric module employed in space applications is the one used to power the US Navy's TRANSIT navigation satellite. This is a twelve-sided structure as shown in Figure 8.6;[9] each side panel is in effect a thermoelectric module consisting of 36 lead telluride thermocouples operating at a hot side temperature of 700 K and arranged in a series–parallel configuration. Each panel produces ~3.2 W at 0.5 V. The thermocouples are made from 2N-PbTe (*n*-elements) and 3P-PbSnMnTe (*p*-elements). A natural development of this arrangement is the 'ring converter' in which the modules take up a ring configuration.[10] Originally conceived to allow the use of the heat rejected from a Si-Ge multihundred watt generator by locating another generator around it, the ring converter was later developed as an independent generator; it has *p*-type thermocouple elements composed of segmented legs of 3P-PbSnMnTe on the hot side and BiSbTe on the cold side. Using selenide thermocouple material this arrangement was employed in the high performance selenide generators.[11]

8.6 D.C. TO D.C. CONVERTER

Conventionally, thermoelectric power is generated at relatively high values of current and low voltage; an increase in voltage is effected by employing a series

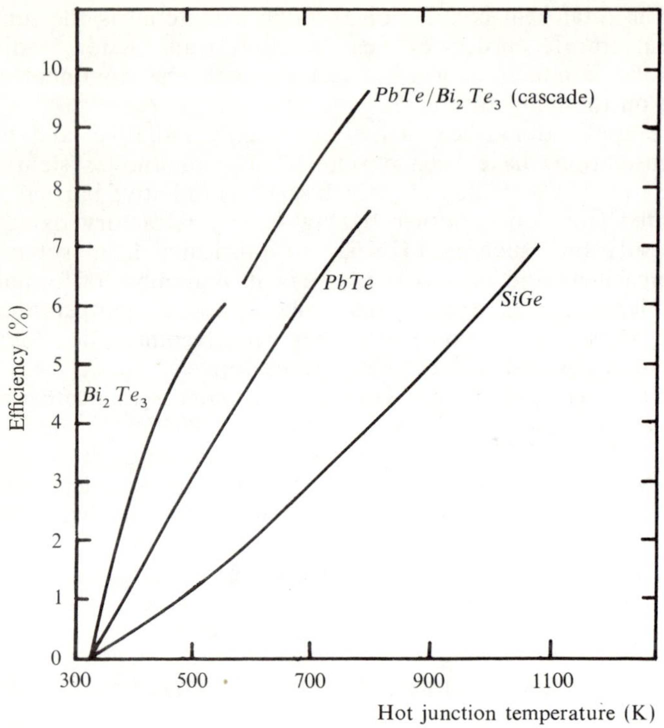

Figure 8.5 Thermoelectric module efficiency. (After Carney, reference 3.)

connection to a multiplicity of thermocouples, with the load voltage from a thermoelectric module being about half the open circuit voltage V_{oc} which is given by $V_{oc} = n\alpha_{ab}(T_{hot} - T_{cold})$, where n is the number of couples connected in series. At high power levels it is possible to obtain high voltages directly by increasing the number of couples. In biomedical applications, however, the desirable reduction in power that accompanies the miniaturization of the device is accompanied by a low voltage output. With most devices requiring input voltages of the order of volts, d.c. to d.c. converters are employed. The efficiency of voltage conversion is independent of power input and is primarily a function of voltage input from the thermoelectric module. A typical performance characteristic is shown in Figure 8.7.[3] Ideally, a sufficient number of thermocouples should be employed to give an output of one volt or so, but an acceptable compromise is often to operate well below 1 V where an efficiency of 60 per cent is still obtainable.

The overall conversion efficiency of a thermoelectric device depends upon its temperature of operation and the proportion of available heat which passes through the thermoelements. In miniature RTGs, the ratio of the thermoelement cross-section area to the surface area of the fuel capsule decreases with reduction in power level. Consequently, the efficiency of the device decreases with a decrease in power level, as shown schematically in Figure 8.8.[12] In addition, for fixed cold side temperature the conversion efficiency optimizes at a different hot side temperature for different power levels, and because the

Figure 8.6 The US Navy's F-1 TRANSIT converter. (After Bradshaw and Postula, reference 9.)

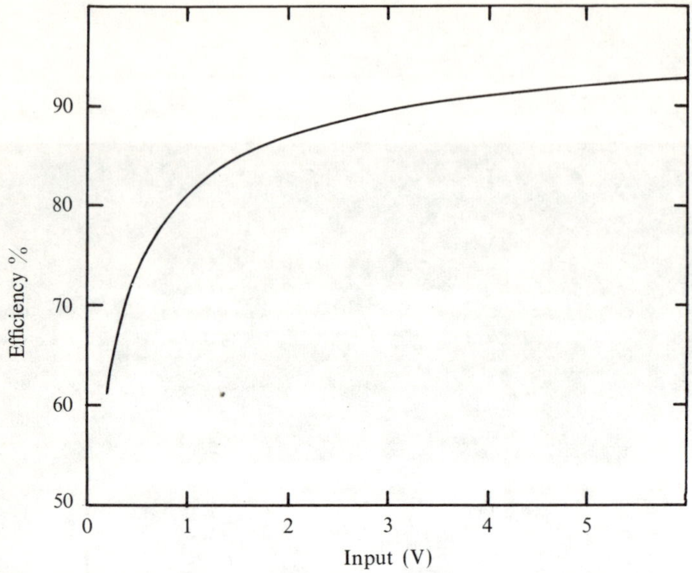

Figure 8.7 D.C. to d.c. voltage converter efficiency. (After Carney, reference 3.)

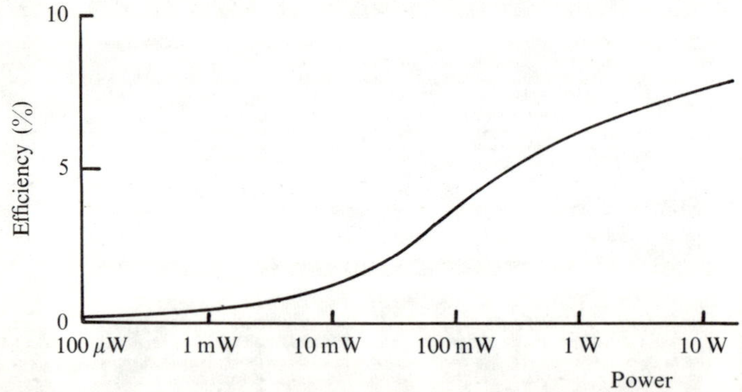

Figure 8.8 Dependence of micro/milliwatt RTG efficiency on power level. (After Raag, reference 12.)

relative amount of thermal insulation decreases with increase in power level, so does the optimum hot side operating temperature.

The production of low power levels at relatively high voltages required in miniature generators can be achieved by using a large number of thermocouples with extreme ratio of length to cross-sectional area. Thermoelectric materials do not normally lend themselves to machining, although it has been reported that the machining characteristics of both Bi_2Te_3 and SiGe have been modified towards this end. The tensile and compressive strength of the tellurides are significantly less than those of the silicon–germanium alloys, and silicon–germanium thermoelements can be machined as little as 125 μm in section compared to about 750 μm for the tellurides. This capability has a

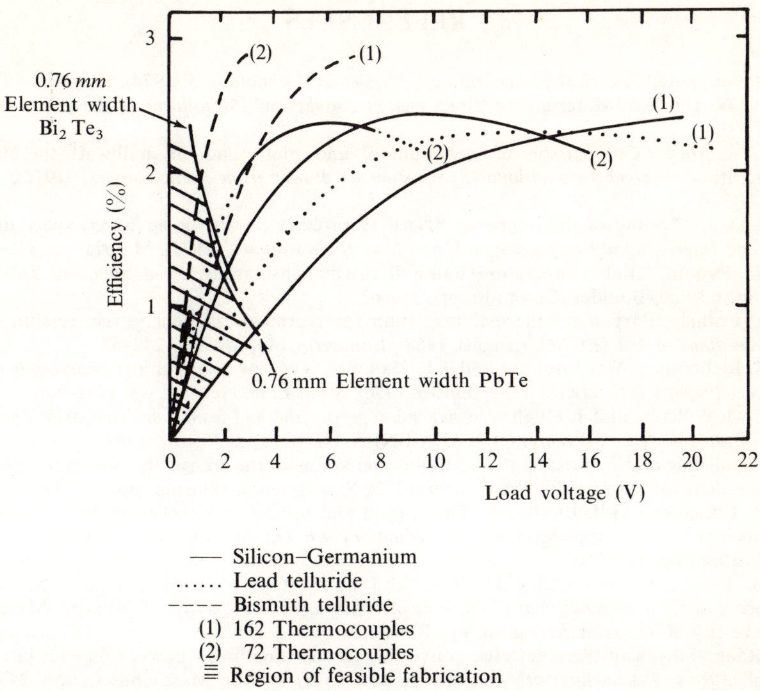

Figure 8.9 Performance analysis of an RTG designed to produce 65 mW of electrical power. (After Raag, reference 12.)

significant effect on the d.c. output of the module and enables RTGs based on the silicon–germanium alloys to produce higher d.c. output voltages than those based on other technologies.[12] Figure 8.9[12] displays the results of the performance analysis of a fibrous insulated RTG designed to produce 65 milliwatts of electrical power. Comparison is made between silicon–germanium, lead telluride and bismuth telluride.

An alternative method of achieving these extremes of geometry is to use wires, films or tapes. Although wire thermoelements, such as those based upon Cupron–Tophel, facilitate module fabrication and deliver high voltages at low power, their conversion efficiency is very low. Thermoelectric materials have been deposited as thin films, usually 1 μm thick or less,[13-18] and thin film thermoelements have been deposited on glass slides in single layers with the number of elements being limited by the size of the substrate.[1] Thermoelements have also been stacked one upon another, separated by insulating SiO_2.[19] Attempts have also been made to deposit thousands of thermocouple cells in a very small volume, although problems associated with pin-holing and excess heating during the fabrication process have prevented multicell fabrication.[20] Siemens of Germany have developed a conversion system consisting of thin film thermocouples deposited by a flash evaporation technique using a tape of polyamide as a substrate.[21] Two tapes are connected in parallel to ensure reliability, and upwards of 700 thermocouples of $Bi_2Te_3/ZnSb$ give 3 to 6 volts under load.

REFERENCES

1. A. Penn, 'Small electrical power sources', *Physics in Technology*, **5** (1974), 115–40.
2. D.A. Wright, See 'Materials for direct energy conversion', *Metallurgical Reviews*, **148** (1970), 147–60.
3. H.C. Carney, 'Comparison of strontium-90 and plutonium-238 milliwatt thermoelectric generators', *Second International Symposium on Power from Radioisotopes*, (1972), Madrid, pp. 143–63.
4. P.J. Dick, 'Testing of the improved SNAP-19 primary power for advanced space missions', *Second International Symposium on Power from Radioisotopes*, (1972), Madrid, pp. 571–88.
5. M.L. Paquin, 'High temperature multifoil thermal insulation', *Proceedings of 3rd IECEC*, (August 1968), Boulder, Colorado, pp. 256–62.
6. J.O. Collins, 'Particulate thermal insulation for thermoelectric energy conversion devices', *Proceedings of 3rd IECEC*, (August 1968), Boulder, Colorado, pp. 249–52.
7. A.R. Lieberman, W.E. Osmeyer and T.E. Hammel, 'The low cost high performance generator', *Proceedings of 11th IECEC*, (September 1976), State Line, Nevada, pp. 1591–8.
8. J.A. McWilliams and J. Hughes, *A new microporous thermal insulation: Industrial applications for isotopic powered generators*, UKAEA-ENEA, Harwell, pp. 363–82, 1969.
9. G. Bradshaw and F. Postula, 'Description and status of the transit thermoelectric generator', *Proceedings of 7th IECEC*, (September 1972), San Diego, California, pp. 200–15.
10. N.B. Elsner and G.B. Bradshaw, 'Fabrication and testing of a thermoelectric ring converter segment for space applications', *Proceedings of 9th IECEC*, (August 1974), San Francisco, California, pp. 165–70.
11. N.B. Elsner, J. Chin and G.H. Reynolds, 'Fabrication of selenide segmented elements', *Proceedings 3rd International Conference on Thermoelectric Energy Conversion*, (March 1980), University of Texas at Arlington, pp. 105–8.
12. V. Raag, 'Review of thermoelectric conversion in micro/milliwatt power range for bio-medical applications', *Proceedings 6th IECEC*, (August 1971), Boston Massachusetts, pp. 245–55.
13. G. Abowitz et al., 'Thin film techniques in thermoelectricity', *Final Report, Contract No.* 88212, Ser. 1734A–984, November 1963.
14. G. Abowitz et al., 'Thin film thermoelectric', *Semiconductor Products and Solid State Technology*, **8**(2) (1965).
15. V.A. Nayer and U.V. Kravchuk, 'Apparatus for producing films of thermoelectric materials', *Kholodil'naya tekhnika i tekhnologiya*, No **10** (1970), 44–8.
16. U.A. Arifov et al., 'Variation in the thermoelectric properties of solid solutions $Bi_2Te_{2.4}Se_{0.6}$ in the course of vacuum deposition', *Geliotekhnika*, **10**(4) (1974), 3–6.
17. U.A. Arifov et al., 'Thermoelectric properties and the structure of films obtained by thermal evaporation of a solid solution based on the tellurides of bismuth and antimony', *Geliotekhnika*, **10**(4) (1974), 62–5.
18. D.K. Benson and C.E. Tracy, 'Design and fabrication of thin film thermoelectric generators', *Proceedings of Fourth International Conference on Thermoelectric Energy Conversion*, (1982), University of Texas at Arlington, 11–14.
19. W. Thorpe, 'Manufacture of Thermoelectric Generators', Patent No. 3, 434, 203, 1969.
20. T. Renner et al., 'An isotopic thermoelectric battery for cardiac pacemakers', *Proceedings of Second International Symposium on Power from Radioisotopes*, (1972), Madrid, p. 207–16.
21. R.K. Collier and C.E. Backus, 'Thin film thermoelectric generators' *Proceedings of 7th IECEC*, (September 1972), San Diego, California, pp. 216–20.

9

Generator Applications

9.1 INTRODUCTION

THERMOELECTRIC generators have no moving parts, are reliable and silent, and can operate unattended in inaccessible and hostile environments. Applications to date, however, are almost exclusive to situations where combinations of the generators' desirable features outweigh the low efficiency and relatively high cost. In principle, provided a temperature difference can be maintained across the thermoelements of a generator, it will deliver power. Applications can conveniently be divided into non-isotope and isotope powered categories.

9.2 NON-ISOTOPE POWERED APPLICATIONS

Thermoelectric generators using fossil fuel heat sources have found application in remote situations where conventional sources of electrical power are not available. Perhaps the earliest examples are from the Soviet Union, and these range from kerosene-powered devices delivering a few watts supply for radio receivers to wood-burning generators supplying up to 500 watts.[1] Efficient utilization of available heat presents problems in burner design, and multi-solid-fuelled burners have been developed in the USA for use with locally available fuel sources.[2] Such fossil-fuelled generators have found military applications (Figure 9.1) where their multifuel capability and silent operation in tactical situations prove very desirable features.[3,4] In these devices a significant reduction in fuel consumption can be achieved by using a regenerative system in which the primary air for combustion is preheated.[5] The capability of military thermoelectric generators is being extended to provide air-conditioned accommodation for combat personnel.[6] Feasibility studies have also indicated that aerodynamically heated thermoelectric converters can be effectively applied to projectiles for powering radio-frequency proximity fuses.[7]

A number of attempts have been made to use solar radiation as a heat source.[8] High temperatures can be obtained by a variety of systems[9,10,11] and a high power thermoelectric generator using a solar furnace has been described.[12] Generators can be constructed using cheap and readily available raw materials[13,14] and the necessity to grow expensive single crystals can be avoided without essentially impairing the efficiency of the thermoelement.

Figure 9.1 500 W military thermoelectric generator. (After Guazzoni, reference 4, courtesy Army Electronics Technology and Devices Laboratory, ERADOM Fort Monmouth, NJ, USA.)

Successful operation is critically dependent on climatic conditions, and although this problem does not arise when orbiting the earth[15] the requirement remains of storing sufficient heat to power the device for 35 minutes of every 90 minutes orbit.[16]

Significant improvements to the performance of a solar-powered thermoelectric generator (SPTG) may be obtained using high temperature thermoelectric materials with the hot junction coated with high temperature solar-absorbing films,[17] for example Mo–Al_2O_2 (stable up to 1300 K[18]) and Ge–Al_2O_2 (stable up to 750 K[19]). An SPTG using selective absorbent coatings and with thermoelements fabricated from fine grained silicon–germanium alloys is predicted to operate with an efficiency of better than 12 per cent between room temperature and 1000 K.[20] Thermoelectric generators are relatively insensitive to radiation damage, and systems based upon SPTGs for providing auxiliary power to spacecraft have been described.[21]

Of the presently available thermoelectric technologies, SPTGs using multiple high voltage thermoelements fabricated from silicon–germanium alloy and with individual solar concentrators offer the optimum combination of technology and configuration for minimizing power source mass. These

SPTGs are currently being examined for possible application in orbit around Mercury.[22]

A novel proposal has been made to overcome the energy storage problem encountered in the development of high altitude ($\sim$22 km) platforms for military and commercial surveillance over distances of up to 600 km. The intention is to utilize the earth's thermal albedo (300 K at 10 μm, $>$140 watts/m^2) as a source and the radiation in the direction of outer space (3.5 K) as a sink. Prototype 'albedo arrays' have been constructed from 1.3 mm copper clad on 2mm 'MYLAR' with bismuth telluride thermocouples laminated laterally between the collectors and radiators to minimize thermal leakage.[23]

A portable flame-powered generator weighing 5.45 kg, using leaded petrol as fuel, has been developed by the General Instrument Co., in the USA.[24] It consumes 2.27 kg of fuel in 12 h, giving 45 watt at 6 V d.c., an overall efficiency of 2.1 per cent. Thomson-CSF, France, market a range of flame-powered generators capable of operating at temperatures of up to 600, 900 and 1300 K,[25] depending on the choice of Bi_2Te_3 or PbTe or SiGe thermoelements. These are available in two models, identified as TG50 which gives 50 W at 12 V, and TG100 which gives 100 W (Figure 9.2). These devices have found applications in radio and television links, repeater stations, automatic meteorological stations, and in the cathodic protection of gas and oil pipes.[25] The possibility of using waste exhaust heat in automobiles to satisfy electrical requirements has been examined.[26,27] Conventional batteries would be retained to provide electrical supply during the buildup to the required exhaust gas temperatures. The idea has been extended to consideration of large scale recovery of thermal waste from electric power plants[28] using sources such as stack flue pipes, gas turbines and condenser steam. However, the conclusions reached have not been encouraging.

Figure 9.2 50W and 100W flame-powered thermoelectric generators, (After Stoll, reference 25; Courtesy Thomson–CSF.)

Studies into the utilization of oceanic temperature gradients (a 15 to 20 K temperature difference is encountered over relatively short changes of depth in the hot brine phenomenon of the central Red Sea)[29] has led to a conceptual design of megawatt thermoelectric units. In off-shore situations generators have also been designed that operate on the temperature difference between seawater and the hot oil as it emerges from the ground.[30] The possibilities of obtaining large scale economic thermoelectric power from low grade sources of heat such as ocean gradients, geothermal and solar ponds have been explored.[31] Although technically feasible, device cost is a major obstruction, and in order to achieve an electrical power cost of less than \$0.1/kWh new generator design and mass production methods must be developed.

The only device that has enjoyed widespread domestic application is the fail-safe gas furnace pilot light.[32] With the jet burner lit, the thermoelectric device delivers enough power to an electromagnet to hold the main gas valve open. This closes if the pilot light goes out.

9.3 ISOTOPE POWERED GENERATORS

The concept of employing isotope energy as a source of heat has resulted in a proliferation in the application of thermoelectric generators. The long life of the heat source coupled with its high energy density facilitates the production of thermoelectric generators that can operate unattended over long periods. Initially developed to provide onboard power in space vehicles, radioisotope powered generators have since found application in a variety of terrestrial, marine and medical situations.

9.3.1 Medical Applications

Traditionally, power for implanted medical devices has been provided by mercury–zinc cells. Although relatively cheap at £200 each, the battery life is short, necessitating replacement by surgery every 2 to 3 years. Nickel–cadmium cells have also been adapted for continuous use by repeated charging by induction from an external power source while the device is in the patient; charged weekly, such batteries provide 2000 charge–discharge cycles, and, taking into account associated circuitry, the effective life is estimated at 15 years.[33,34] More recently, lithium halide cells have also been developed. Although it has yet to be demonstrated, they are designed to have an operating lifetime of 7 to 10 years.[35] This is a short time on a human lifetime scale, and clearly there is a requirement, especially in younger people, to extend considerably the lifetime of the power source.

The development of miniature nuclear-powered batteries in the early 1970s provided physicians and surgeons for the first time with power sources that contained sufficient energy to drive implantable devices for their intended period of operation. Medical applications can be categorized into two groups: energetic devices with power output of the order of watts, and cybernetic devices which require very little power ($<300\,\mu$W) and are essentially stimulators.

As far as energetic devices are concerned it seems unlikely that thermoelectric devices will make a significant contribution, and research is currently concentrated on developing a variety of nuclear-powered heat engines for ventricular assist systems[36] and totally implantable hearts.[37] The estimated power requirements listed in Table 9.1 are based upon the energy requirements of the biological process itself. In practice a cardiac pacemaker, for example, requires about 300 μW; most of the electrical power, about 90 per cent, is wasted at the electrode/tissue interface.

Table 9.1 *Power requirements of medical applications*

Energetic devices (Power output, W)	Cybernetic devices (Power output, mW)
cardiac assist pumps	cardiac pacemaker ~3 μW
artificial heart 3 W	caratoid sinus nerve (and pump) 300 μW
lungs 5 W	sphincter muscle 1 mW
kidneys 5 W	glucose concentration 100 mW
	phrenic nerve

The control tasks performed by cybernetic devices can hardly be described in terms of power output. The devices almost exclusively serve to apply stimulating currents to nerves or muscles, such as the phrenic nerve in the diaphragm stimulator and in the treatment of angina pectoris and incontinence. Successful treatment of chronic pain has historically been elusive, but some success has been reported through stimulations of the sciatic nerves.[38,39]

The radioisotope cardiac pacemaker (RCP) is certainly the best-known application of miniature nuclear-powered thermoelectric sources. The pacemaker controls heart rhythms by delivering electrical impulses to the muscle of the heart. The first implant took place at Hospital Broussais, Paris, in 1970 and was an RCP manufactured by Medronic.[40] In essence, the RCP consists of an isotope powered thermoelectric battery using Pu-238 as a heat source, bismuth-telluride thermoelements, a d.c.–d.c. converter and a pulse generator. Typically, the pulse generator is inhibited by normal sinus rhythm down to around 62 beats per minute, but during a heart block the unit paces at a rate of 72 beats per minute.[41] Plutonium has a half-life of 87 years and so the thermal output from the fuel capsule only decreases by 4 per cent in 10 years, and approximating for possible losses in thermal insulation and degradation in thermoelectric material performance, the energy should fall by 60 μJ during this period. In practice, there may be a decrease in output voltage as a result of loss of thermal insulation. The pacemaker is hermetically sealed in a titanium case or encapsulated in an epoxy resin that is compatible with living tissue, and is normally located inside the chest.[40] By 1972 the number of implants had risen to 30, and by September 1975 stood at 1600. About 150 000 heart pacemakers of all types were sold in 1976, and if only 2 per cent of these were nuclear powered at £2000 a time this represents a multimillion-pound business. Competition among the relatively few vendors is keen, with the largest share of the market going to Medronic and the second largest share being taken by Cordis. The UK's activity in this area centres on a battery developed by UKAEA, Harwell (Figure 9.3) and marketed through Devices Ltd. About one hundred preproduction devices have been inserted over a long period at recognized cardiological centres under DHSS supervision. The device pacemaker with AERE nuclear battery is shown in Figure 9.4. Both the tantalum

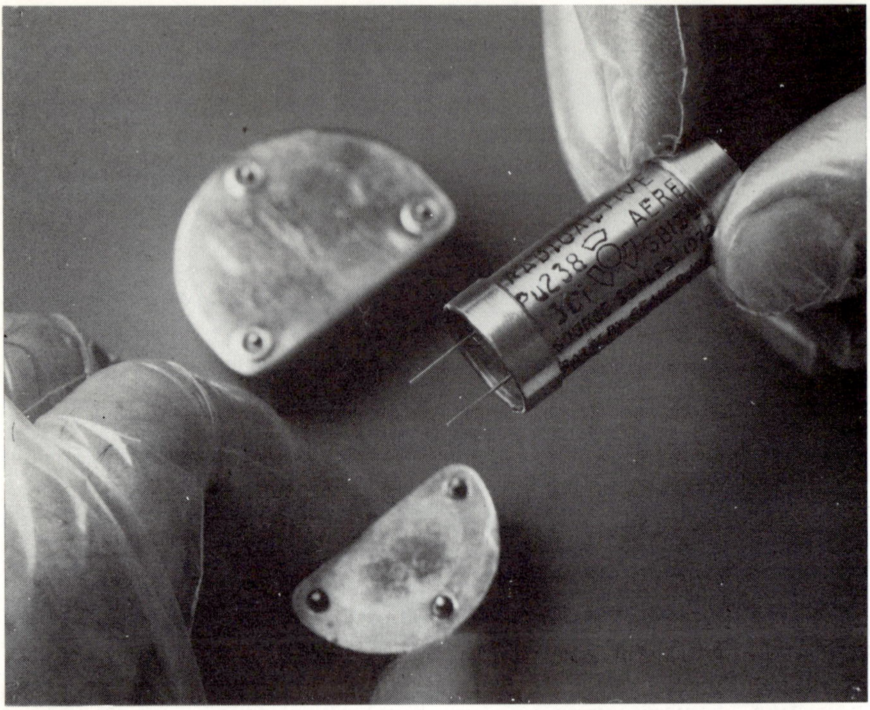

Figure 9.3 AERE miniature nuclear battery. (Courtesy United Kingdom Atomic Energy Authority, Harwell, UK.)

10 per cent tungsten capsule, which can withstand cremation temperatures up to 1600 K, and the battery case are hermetically sealed and effectively serve as primary and secondary containment for the radioactive fuel. Any loss of containment will manifest itself in a noticeable change in battery performance. Although the radiation level at the surface of the device is approximately 30 μGy/h, i.e., less than that at the surface of a luminous watch, plutonium-238 and its decay product Pu-236 are very carcinogenic, and very strict regulations concerning the safety standards have been set by the European Nuclear Energy Agency (ENEA). This agency groups eighteen European member countries, Japan, the USA and Canada. The safety standards include containment of the source to withstand 1000 kg crush tests carried out at various orientations and cremation temperatures of 1600 K for 90 minutes.

Conflicting views have been expressed as to whether or not nuclear pacemakers are really necessary.[41,42] Vendors of devices, together with the patient recipients, see considerable advantages in nuclear-powered thermoelectric batteries. Although the market for miniature RTGs is not booming, a stable market is forecast[43] with a benefit to the patient of considerable increase in the time interval between operations. However, a majority of physicians, responding to a questionnaire from the US Nuclear Regulatory Commission, were of the opinion that nuclear pacemakers are too costly, involve too much paperwork, and have very restricted application. Pacesetter Systems Inc., which markets a nuclear battery, conducted a survey in 1974, and reported that most

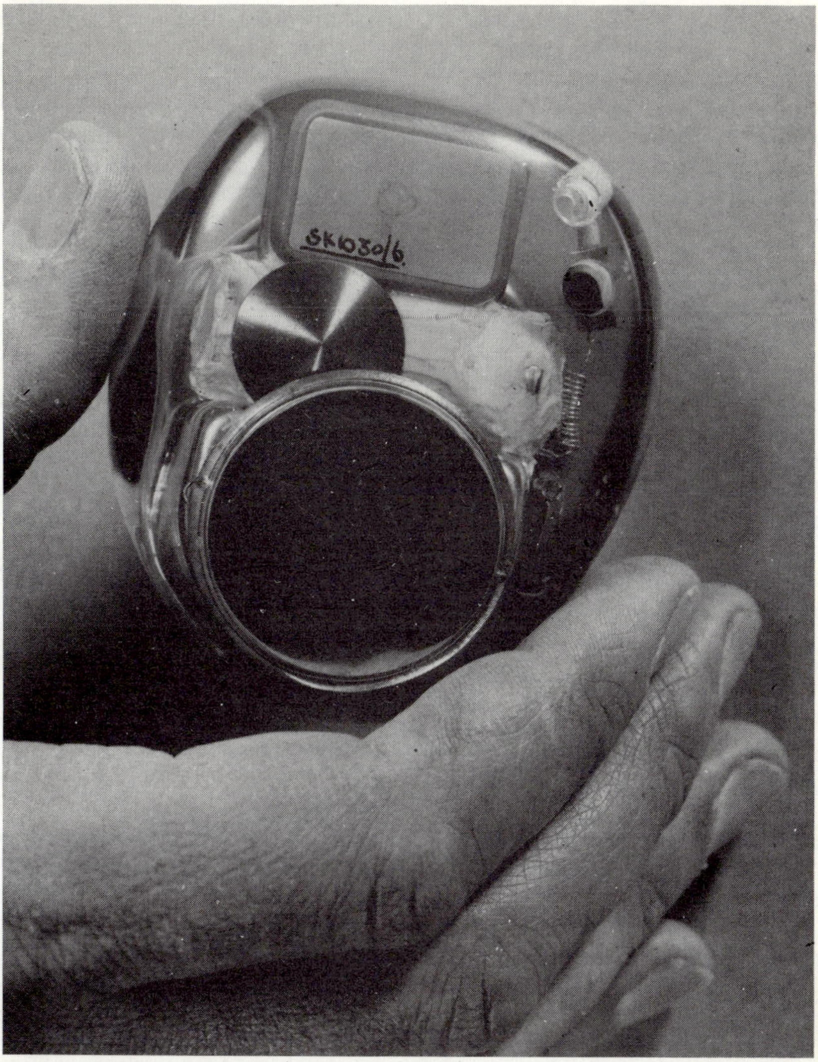

Figure 9.4 Radioisotope-powered thermoelectric cardiac pacemaker. (Courtesy United Kingdom Atomic Energy Authority, Harwell, UK.)

comments were disapproving on account of the difficulties with disposal of the source if the wearer is killed in an accident, and problems associated with obligations to retrieve the unit after the patient's death.[44]

At present, there are no proven alternatives to nuclear-powered pacemakers if one is aiming at avoiding the risks involved in reimplanting pacemakers. In addition, any such alternatives as there are involve implanting in the body large quantities of mercury (which is highly toxic) or lithium (which reacts violently if it comes in contact with body fluids). Also, these batteries are not encapsulated in a way that would prevent the release of the contents in the event of cremation.

Initially in the UK, only selected hospitals were issued with a special-materials licence that specified the number of nuclear-powered pacemaker implants that could be made per year. In the USA the Nuclear Regulatory Commission (NRC) licensed manufacturers to sell a specific number. However, the NRC, in its Final Generic Environmental Statement on the routine use of plutonium-powered cardiac pacemakers, concluded that, based on a balance of the benefits and risks involved, plutonium-powered pacemakers can be licensed for routine use.[45]

9.3.2 Marine and Terrestrial Applications

Accompanying the exploitation of the earth's resources in increasingly remote situations is a requirement for autonomous energy sources capable of long-term operation with a minimal maintenance. Isotope-powered generators are ideally suited to such applications.

Marine applications often introduce special constraints on instrument energy requirements, which provide RTGs with some advantage; RTGs are competitive with, or even superior to, other systems for missions longer than six months and power levels up to about 50 kW. Figure 9.5 shows that they have a potential for widespread use in data buoys, life support systems, submersible habitats and for locating and navigational aids.[46] The weight and cost, based upon 1970 estimates (Figure 9.6), are competitive with other energy sources;[47] the relatively high price is a reflection of the radioisotope costs and presents the greatest obstacle to broader applications. The US Navy is easily the largest operator of marine isotopic-powered generators; typical generators employed by them are Gulf Millibats,[48] designed for operating at depths of up to 10 km. Bismuth telluride thermoelements driven by a strontium-90 heat source deliver at least 1 W of power at 1.5 V to 1.8 V for a ten-year operational period. The output voltage is d.c.–d.c. converted to the required 24 V. In 1977 two Millibats were sited on the floor of the South Atlantic with the intention of coupling them to the deep ocean transponder system incorporated in the American missile impact locating arrays.[49]

The US Navy's requirement for an even longer period of operation has been met by a half-watt RTG powered by plutonium-238.[50] This is designed to deliver a minimum of 0.5 W at 6.0 to 8.5 V by the end of a 15-year mission. (This is equivalent to 65.7 kWh of electricity.) Initially intended for use in cables in undersea surveillance systems, the battery has other potential applications where power sources of high reliability are required, such as in fibre-optic cables.

German activity in submersible RTGs has centred around the TRISTAN generator.[51] Employing partially cascaded thermoelements of Bi_2Te_3 and Si–Ge, it produced 5.4 V and 4.5 A and via the d.c.–d.c. converter 2 W at 24 V.

United States surface and terrestrial applications range from establishing the first SNAP system for nuclear power at Curtis Bay in the Chesapeake on 14 December 1961[52] to an automatic weather station at Umeat in Alaska.[53] The station, with an unattended operating life of at least six years, transmits weather information to aircraft associated with the construction of the trans-

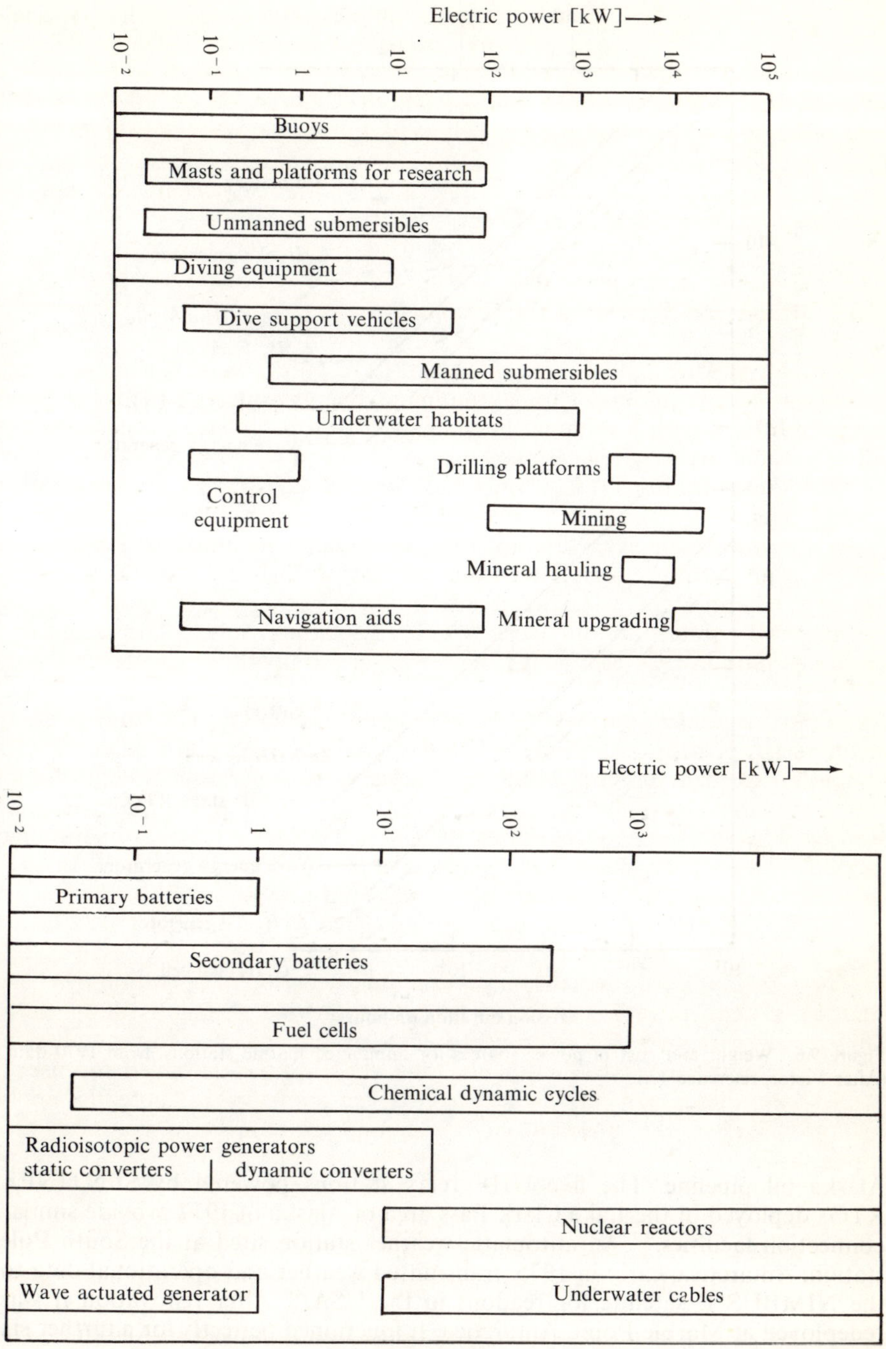

Figure 9.5 Power ranges of various maritime equipment and power sources. (After Victor, reference 46.)

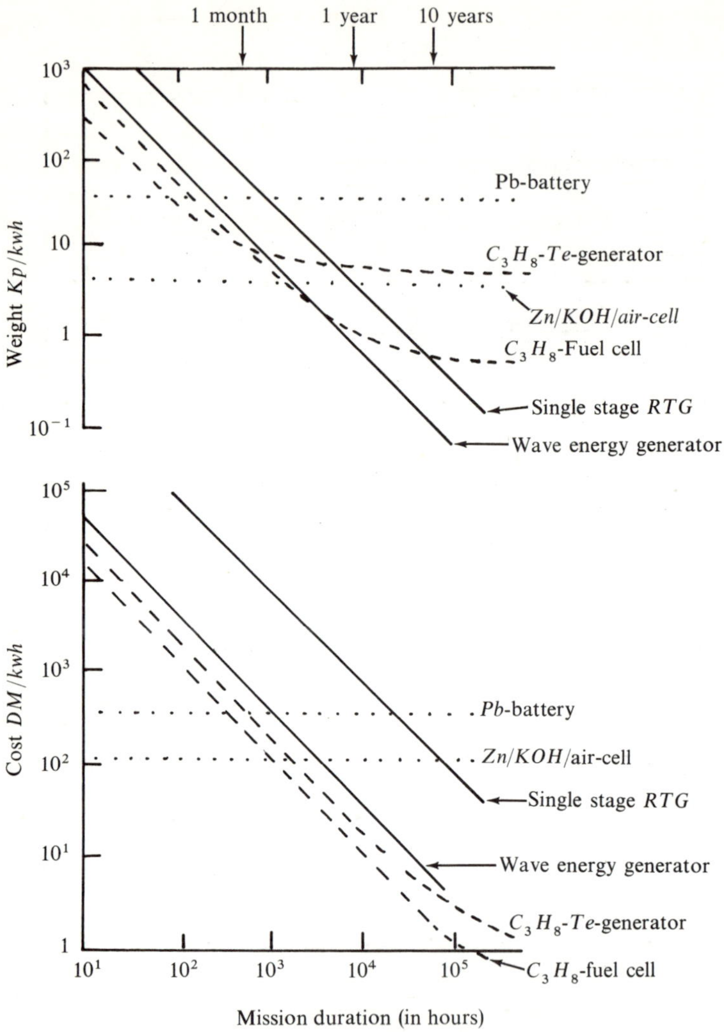

Figure 9.6 Weight and cost of power sources for unmanned marine stations, from 1970 data. (After Victor, reference 47.)

Alaska oil pipeline. The five UHF relay stations powered by SENTINEL RTGs deployed in the Lake Clark Pass area of Alaska in 1977 provide similar connection facilities.[49] An automatic weather station sited at the South Pole station, Antarctica, early in 1975, transmitted weather and operational data to the NIMBUS F satellite for readout in the USA.[49] After renovation it was redeployed at Marble Point, Antarctica. It functioned perfectly for a further six months until an electronic malfunction occurred in the transmitter's integrated circuits.

France's first RTG was commissioned in July 1966. Effort was concentrated on producing an autonomous source of power for off-shore experiments, and by August 1968 a MARGUERITE generator was operating in 75 m of water in the Mediterranean. Applications now range from miniature batteries (GIPSIE) for the cardiac pacemakers to generators providing power to isolated geophysical stations.[54]

The UKAEA have been successful in their RIPPLE (radio isotope powered prolonged life equipment) series of generators.[55] The programme was initiated in 1963, and the first field trial was carried out in 1967. Systems reliability is attested to by the fact that none of the field-trial generators has required attention during mission life, which in two cases exceeded eight years. Early RIPPLE generators, designated I to IX, are in the low-power range. RIPPLE-X is based upon a modular design concept. Up to six multiples of a basic 8 W(e) unit can be connected together, providing generators with outputs in the range 8 to 48 W. A RIPPLE-X generator, purchased by Trinity House and installed in a lighthouse in June 1974, is shown in Figure 9.7.[56] The requirement was for a power output of at least 25 W over a five-year period, and a four-unit system was designed giving start-of-operation power of 32 W. A continuous trickle charge is delivered to a bank of nickel–cadmium batteries, which provides power for the lighthouse tungsten-filament lamps. The radioisotope source used is strontium-90 in its virtually insoluble titanate form. The source is doubly encapsulated in two high-integrity hermetically sealed vessels, which are themselves contained in robust, biological shielding; the generator complies with the provision of relevant national and international safety regulations.

The strontium titanate that fuels all strontium-90 powered thermoelectric generators, other than ones originating in communist countries, has been produced at Oak Ridge, USA, and sold at far below commercial cost to prime the market. The source is now virtually exhausted, and any future supplies will be at a commercial rate. Meanwhile, waste encapsulation and storage-facility capsules are being marketed by the Richland Laboratories at a basic price of 0.25 cent/GBq, but only in one standard size.[57] This delivers approximately 1 kW of heat at the beginning of its life at a basic cost of $15 000. The source geometry is probably better suited for thermomechanical generators than thermoelectric ones.

The Canadian MAPLE (minor atomic prolonged life equipment) generators are unusual inasmuch as cobalt-60 is employed as a heat source.[58] Cobalt is readily available in Canada and has a good energy density and favourable thermal properties. It is, however, potentially at a disadvantage on two accounts. It has a short half-life of 5.26 years, which precludes application in long-lived devices. In addition, most of the decay energy is carried by γ-rays and, to make best use of the heat source, these should be absorbed within as small a volume as possible. This is achieved by locating the uranium or tungsten shielding inside the thermal insulation. The thermoelectric modules are then in contact with the outer surface of the shield. The disadvantage is that the surface energy density of the fuel plus the shielding materials is much less than that of the fuel alone. Consequently, the efficiency of the heat source is greatly reduced even if good thermal insulation is used. However, this shielding arrangement does offer some advantages in that, for a given shield thickness, the mass of shielding is greatly reduced. Also, once the source has been inserted into the shield, little radiation is encountered during the subsequent assembly of the system.

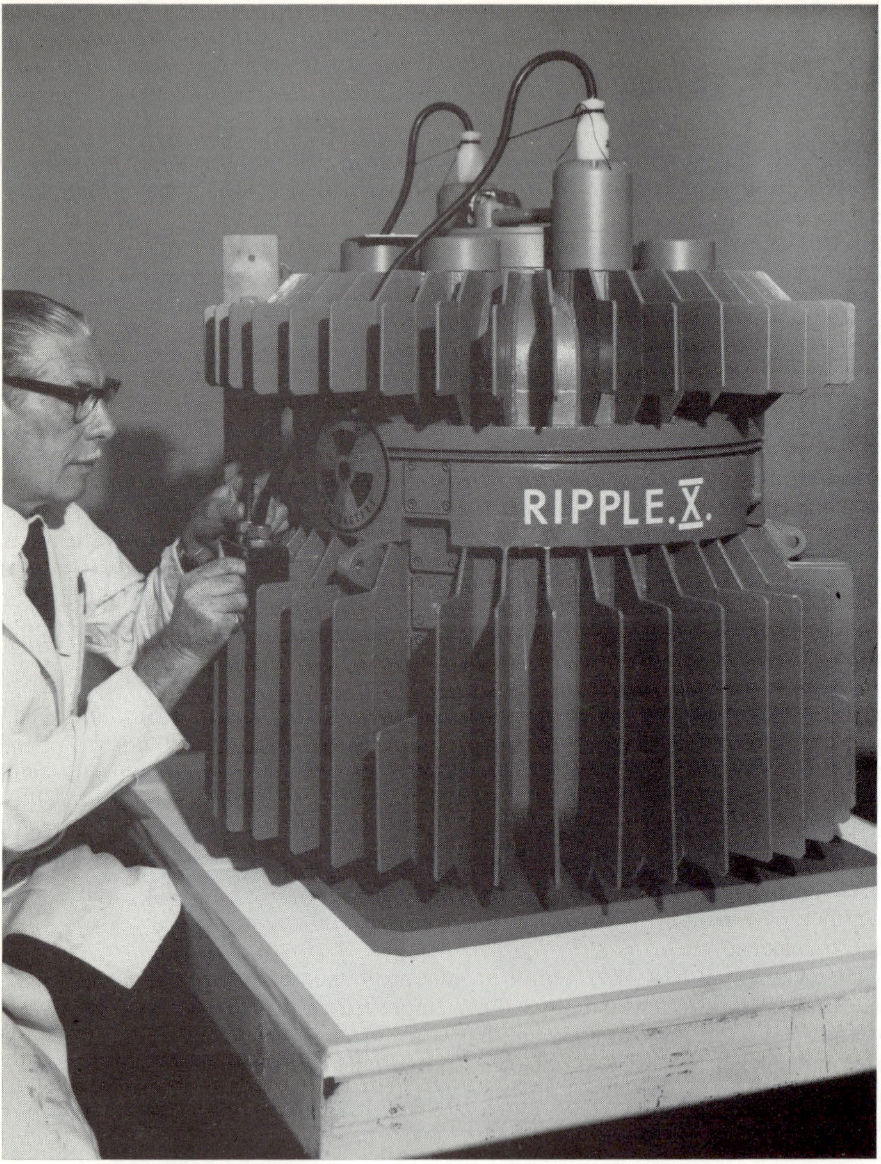

Figure 9.7 RIPPLE-X thermoelectric generator. (After Cooke-Yarborough and Yeats, reference 55, courtesy United Kingdom Atomic Energy Authority, Harwell, UK.)

9.3.3 Space Applications

The development of Systems for Nuclear Auxiliary Power (SNAP) for satellite application was begun in 1955 with the first application of radioisotope power in space accompanying the successful launching and operation of a SNAP-3A

Table 9.2 *Summary of space nuclear power systems launched by USA 1961–1982.*

Flight number	System	Mission	Launch date	Power in watts	Status
Radioisotope power systems					
1	SNAP 3	Transit 4-A	6/61	2.7	Successfully achieved orbit.
2	SNAP 3	Transit 4-B	11/61	2.7	Successfully achieved orbit.
3	SNAP 9A	Transit 5-BN-1	9/63	26.8	Successfully achieved orbit.
4	SNAP 9A	Transit 5-BN-2	12/63	26.8	Successfully achieved orbit.
5	SNAP 9A	Transit 5-BN-3	4/64	26.8	Mission aborted; burned up on re-entry.
6	SNAP 19	Nimbus B-1	5/68	28.1	Mission aborted; heat source retrieved.
7	SNAP 19	Nimbus III	4/69	28.1	Successfully achieved orbit.
8	SNAP 27	Apollo-12	11/69	73.6	Successfully placed on moon.
9	SNAP 27	Apollo-13	4/70	73.6	Mission aborted on way to moon. Heat source returned to south Pacific ocean.
10	SNAP 27	Apollo-14	1/71	73.6	Successfully placed on moon.
11	SNAP 27	Apollo-15	7/71	73.6	Successfully placed on moon.
12	SNAP 27	Apollo-16	4/72	73.6	Successfully placed on moon.
13	SNAP 19	Pioneer F/10	3/72	40.7	Successfully operated to Jupiter.
14	RTG	Transit	9/72	35.6	Successfully achieved orbit.
15	SNAP 27	Apollo-17	12/72	73.6	Successfully placed on moon.
16	SNAP 19	Pioneer G/11	4/73	40.7	Successfully operated to Jupiter. and Saturn.
17	SNAP 19	Viking-1	8/75	42.7	Successfully landed on Mars.
18	SNAP-19	Viking-2	9/75	42.7	Successfully landed on Mars.
19	MHW	LES 8/9	3/76	158	Successfully achieved orbit.
20	MHW	Voyager I	8/77	155	Successfully operated to Jupiter and Saturn.
21	MHW	Voyager II	9/77	155	Successfully operated to Jupiter and Saturn on to Uranus and Neptune.
Reactor power systems (one only)					
22	SNAP 10A	SNAPSHOT	4/3/65	500	Successfully achieved orbit.

(After O'Riordan, reference 69)

Figure 9.8 The TRANSIT satellite; first application of radioisotope power in space. (After Dick, reference 59; courtesy John Hopkins University, USA.)

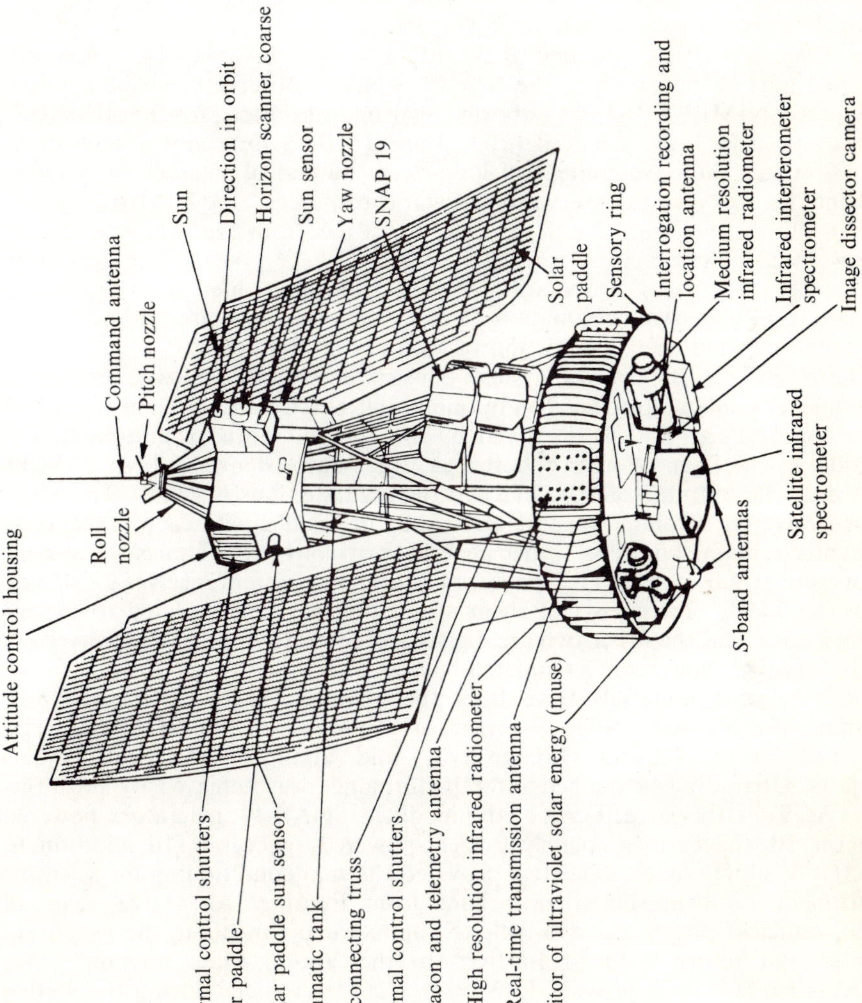

Figure 9.9 The NIMBUS spacecraft. (After Fihelly et al., reference 61.)

generator aboard the US Navy's TRANSIT 4-A satellite in June 1961.[59] (A summary of space nuclear power systems launched by the USA is given in Table 9.2.) Using encapsulated plutonium-238 as a heat source, the SNAP-3A generator is totally enclosed in an external spherical housing about 13 cm in diameter (Figure 9.8). The SNAP-3A led to the development of the generator used to power the US Navy's TRANSIT TRIAD navigational satellite. Launched on 2 September 1972 the generator was designed to operate for more than five years and deliver 35 W at 3 V.[60]

In 1968, SNAP-19, the first RTG to be flown on a NASA spacecraft, supplied part of the power to the NIMBUS NASA observatory. The mission, designated NIMBUS B-1, was aborted two minutes after launch. NIMBUS-III, launched in 1969, achieved orbit. The NIMBUS programme was orientated towards the development of long-range numerical weather predictions and the discovery and conservation of world resources.[61] A NIMBUS spacecraft is shown in Figure 9.9. It receives 24.5 V d.c. from the solar-conversion power supply connected in parallel with a SNAP-19 system. The generator failed after three years of operation. Parallel laboratory life tests[62] attributed the failure to excessive sublimation of the thermoelectric material and loss of hot-junction bond due to depletion of the internal gas cover.

The SNAP-19 B-1 power system consists essentially of two generators mounted in tandem, a power conditioning unit (PCU) and a telemetry signal conditioning unit (TSCU). Powered by 570 thermal watts of plutonium-238 dioxide, in the form of microspheres, each generator delivers about 28 W at 2.6 V d.c. The output from the PCU to the regulated bus is 24.5 V. The generator–converter combinations are electrically in parallel with each other; consequently a major failure of either generator or converter enables the system to operate at half power. The spacecraft's telemetry system receives 28 signals from the TSCU. Twenty-six of them are analogue signals, related to system performance, and the other two are digital and indicate whether the converters are on or off the spacecraft's regulated busbar.

Lead-telluride materials have been the mainstay of the US space programme, the original SNAP-19 generator employed on the NIMBUS being modified for the Pioneer (Jupiter flyby) and Viking[63] (Mars soft lander) projects. Upgrading of the generator performance was achieved by substituting TAGS for the *p*-leg. Four of the modified SNAP-19 generators powered Pioneer 10, the first completely nuclear-powered spacecraft. In addition to the 160 W supply, nuclear heaters provided thermal conditioning for scientific instruments and prevented propellant from freezing. After five years of flight, sufficient power was available to conduct experiments in the vicinity of Jupiter and return data to Earth.[64] In the Viking Mars mission[65] two SNAP-19 RTGs each provide 35 W to the 'lander vehicle' during the 90-day Mars surface mission.

SNAP-27 generators employing PbTe/TAGS thermocouples provide power for the Apollo lunar surface experimental package (ALSEP) Figure 9.10. Their generators are conceptually different from others in that they are designed to be fuelled manually on the surface of the moon.[66] The fuel, in the form of microspheres, is loaded into a graphite lunar module fuel cask (GLFC) one day before the launch. The GLFC is designed to protect the capsule from re-entry heating in the event of an abort. This is exactly what did happen on the ill-fated Apollo 13 mission, and the cask now rests in a deep trench in the

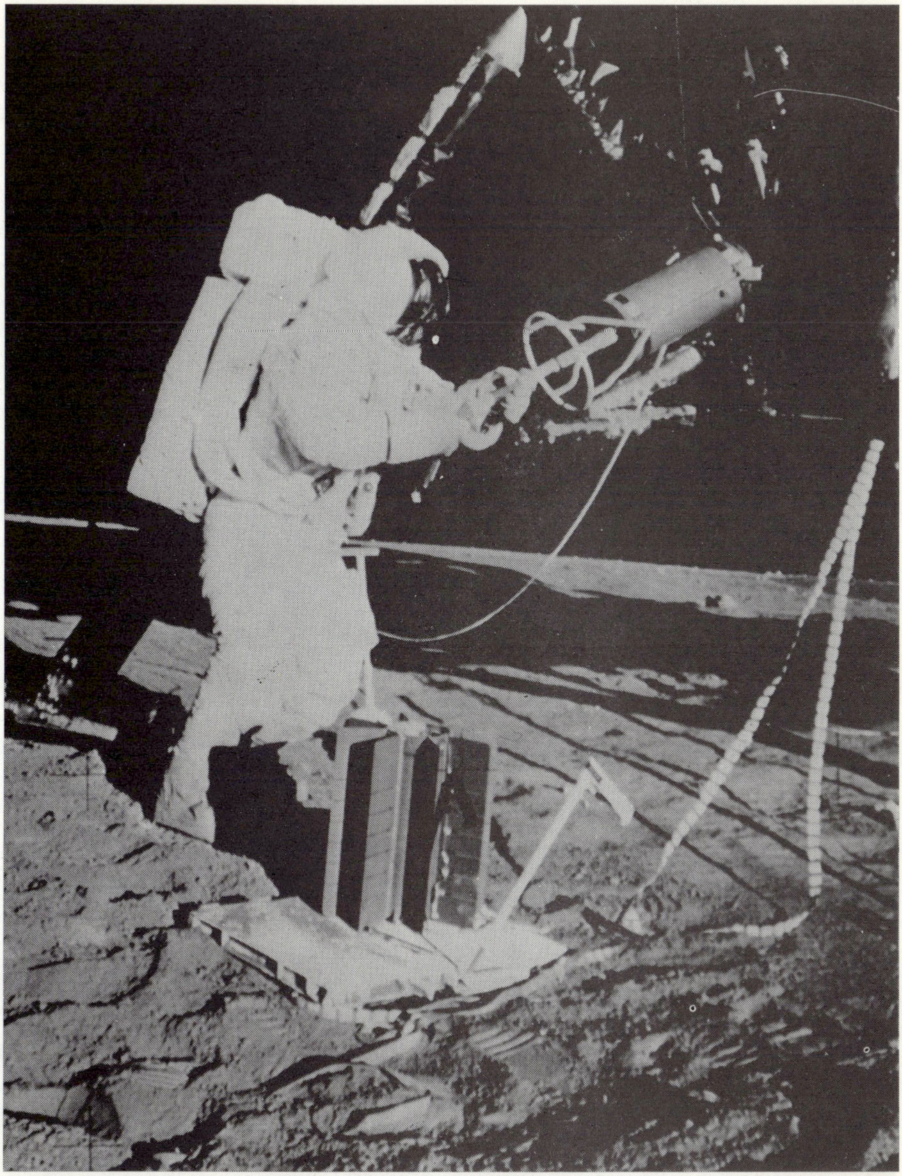

Figure 9.10 Apollo lunar surface experimental package powered by a SNAP-27 thermoelectric generator, (Courtesy NASA and the Jet Propulsion Laboratory.)

South Pacific. To date, five generators have been positioned on the surface of the moon.

Thermoelectric generators based upon silicon–germanium alloys have been used in the multihundred watt (MHW) RTG[67] developed for use in the US Air Force communication satellite launched in 1976. Following a comprehensive material evaluation[68] similar MHW generators were used to power the 1977 Jupiter–Saturn Mariner spacecrafts Voyager 1 and 2 (Figure 9.11). Fairchild

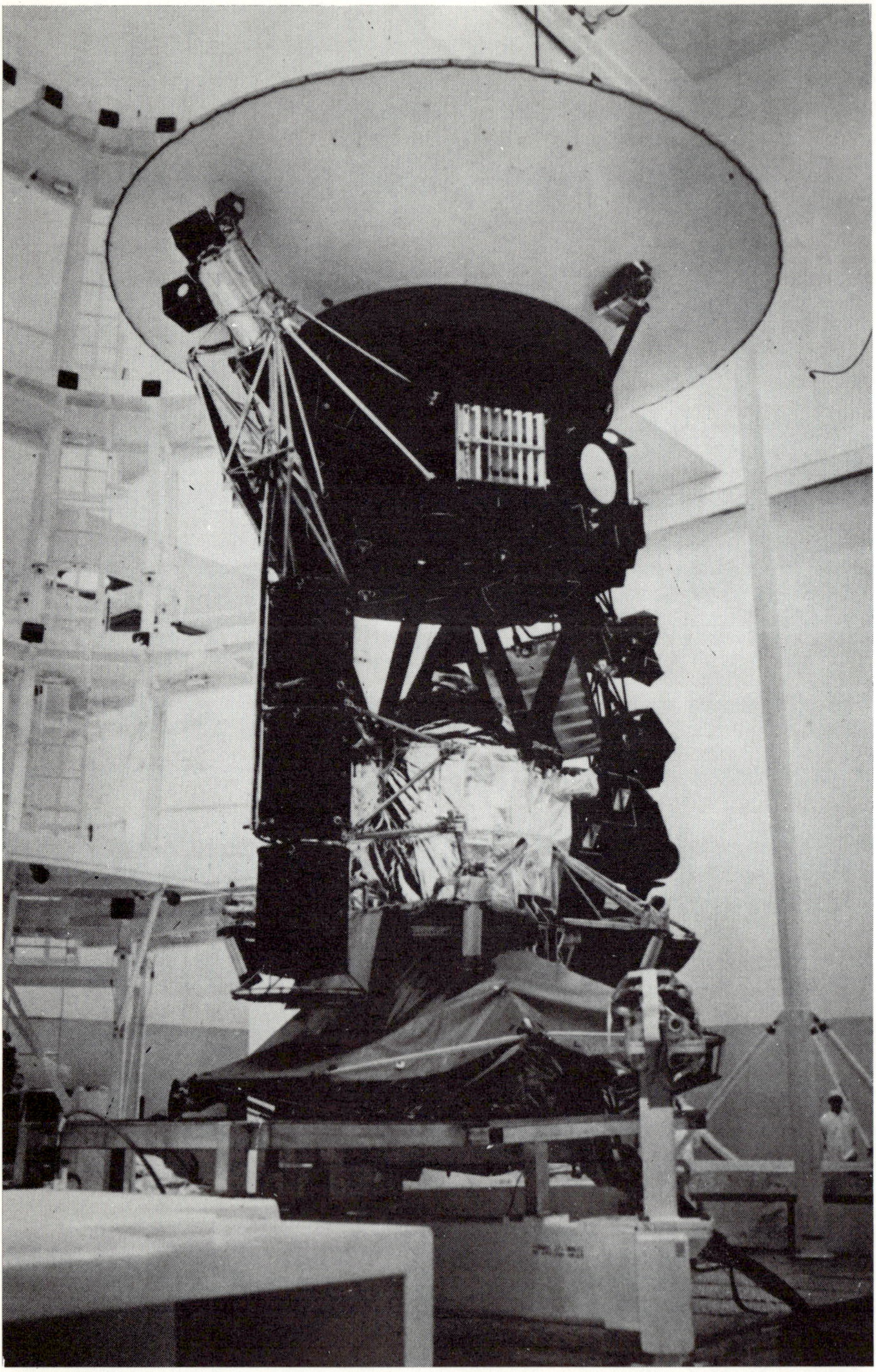

Figure 9.11 The Jupiter–Saturn mariner spacecraft 'Voyager'. (Courtesy Jet Propulsion Laboratory, USA.)

Space and Electronics company have produced an RTG design which incorporates a modular general-purpose heat source, silicon–germanium gallium phosphide thermocouples and an advanced thermoelectric module concept. Design analysis of the generator predicts that a twelve-unit 282 W (e) system will operate at a conversion efficiency of 9.41 per cent.[69]

The development of the use of RTGs in space has been accompanied by an increase in the complexity of the safety requirements, each new space vehicle and mission flight presenting its own unique set of conditions and safety procedures that are incorporated at the early design stage.[70] Systems have been developed that ensure containment through all phases of the mission.[71]

REFERENCES

1. See A.F. Ioffe, *Semiconductor Thermoelements and Thermoelectric Cooling*, Infosearch, London, 1957.
2. H.R. Hazard, 'Multi-solid fuel burner for thermal energy conversion systems', *Proceedings of 17th Annual Power Sources Conference*, (1963), New Jersey, USA, pp. 28–30.
3. J.P. Angello and G.R. Frysinger, 'Military multi-fuel thermoelectric generators', *Proceedings of 3rd IECEC*, (1968), 986–9.
4. G. Guazzoni, '500 watt thermoelectric power source for military use', *Proceedings of First International Conference on Thermoelectric Energy Conversion*, (1976), 136–42.
5. G. Guazzoni, J. Angello and A. Herchakowski, 'Regenerative burner system for thermoelectric and power sources', *Proceedings of Second International Conference on Thermoelectric Energy Conversion*, (1978), 121–5.
6. G. Guazzoni, 'Thermoelectric generators for military applications', *Proceedings of Fourth International Conference on Thermoelectric Energy Conversion*, (1982), University of Texas at Arlington, paper 2 (1).
7. P.E. Eggers, 'Aerodynamically heated thermoelectric converters for powering proximity fuses in 20 mm and 30 mm projectiles', *Proceedings of 10th IECEC*, (1975), 736–49.
8. M. Telkes, 'Solar thermoelectric generators', *Journal of Applied Physics*, **25**(6) (1954), 765–77.
9. A. Arifov, 'Solar thermoelectric generator with cylindrical receiver', *Applied Solar Energy*, **3** (1967), 1–4.
10. M.H. Cobble and P.R. Smith, 'Overall efficiency of a solar collector using a Fresnel lens and thermoelectric module to generate electrical power', *Proceedings of First International Conference on Thermoelectric Energy Conversion*, (1976), 162–7.
11. A. Brandstetter and G. Yekutieli, 'Triangular electrode high power density thermopiles', *Proceedings of First International Conference on Thermoelectric Energy Conversion*, (1976), p. 156.
12. A. Brandstetter and G. Yekutieli, 'Solar furnace type high power density thermoelectric generator', *Proceedings of 13th IECEC*, (1978), 1989–95.
13. K. Landecker, 'Conversion of solar energy into electrical power by means of the thermoelectric effect', *Proceedings of First International Conference on Thermoelectric Energy Conversion*, (1976), 1–6.
14. H.J. Goldsmid and J.V. Cochrane 'Solar thermoelectric generators', *Proceedings of Fourth International Conference on Thermoelectric Energy Conversion*, (1982), University of Texas at Arlington, 7–11. (2).
15. G.W. Glassburn, 'Flat plate solar thermoelectric generator orbital experiment', *IEEE Transactions* AS-**1** (1963), 1396–403.
16. P.E. Kueser et al., 'Experimental results from a solar heated thermoelectric generator utilizing thermal energy storage', *IEEE Transactions* AS-**2** (1964), 675–81.
17. K.P. Suleebka, 'High temperature solar thermoelectric generator', *Applied Energy*, **5**(1) (1979), 53–61.
18. R.N. Schmidt and K.C. Park, 'High temperature, space-stable selective solar absorber coatings', *Applied Optics* **4** (1965), 917–25.

19. J. Gittleman, 'Applications of granular semiconductors to photothermal conversion of solar energy', *Applied Physics Letters* **28** (1976), 370–1.
20. D.M. Rowe, 'A high performance solar powered thermoelectric generator', *Applied Energy*, **8** (1981), 269–73.
21. V. Raag and R.E. Berlin, 'A silicon–germanium solar thermoelectric generator', *Energy Conversion*, **8** (1968), 161–8.
22. V. Raag and L.C. Hankins, 'Design concepts of solar thermoelectric generators in space applications', *Proceedings of Second International Conference on Thermoelectric Energy Conversion*, (March 1978), University of Texas at Arlington, pp. 60–5.
23. L.W. Lemley, 'A radiation thermoelectric power converter', *Proceedings of Third International Conference on Thermoelectric Energy Conversion*, (March 1980), University of Texas at Arlington, pp. 20–7.
24. C. Teleki, '45 Watt thermoelectric generator', *Proceedings of 17th Annual Power Sources Conference*, (1963), New Jersey, USA, PSC Publications Committee, pp. 31–3.
25. R. Stoll, 'Les generateurs thermoelectriques', *Entropie*, **52** (1973), p. 40.
26. B.L. Embry and J.R. Tudor, 'A thermoelectric generator powered by engine exhaust heat', *Proceedings of 3rd IECEC*, (August 1968), Boulder, Colorado, pp. 996–1007.
27. A.W. Serksnis, 'Thermoelectric generator for an automotive charging system', *Proceedings of 11th IECEC*, (September 1976), State Line, Nevada, pp. 1614–18.
28. M.A. Shirazi, 'Thermoelectric generators powered by thermal waste from electric power plants', *Proceedings of 5th IECEC*, (September 1970), Las Vegas, Nevada, pp. 6–17.
29. A. Brandstetter, 'The hot deeps of the Red Sea as a potential heat source for thermoelectric power generation', *Proceedings of 9th IECEC*, (August 1974), San Francisco, California, pp. 277–80.
30. A. Penn, UKAEA, private communication.
31. D.K. Benson and T.S. Jayadev, 'Thermoelectric energy conversion—economic electric power from low grade heat', *Proceedings of Third International Conference on Thermoelectric Energy Conversion*, (March 1980), University of Texas at Arlington, pp. 27–57.
32. 'Fail safe gas pilot light', *Sperrin Data Sheet*, Courtesy of Wales Gas.
33. 'Battery-powered pacemaker allows near permanent use', *Medical Equipment Journal of Japan*, (January 1973). See 'Newsletter' *Information Bulletin on Isotopic Generators and Batteries*, **3**(2) (1973), p. 3.
34. 'You can rely on a rechargeable pacemaker', *New Scientist*, (1975).
35. Are nuclear pacemakers necessary?', *New Scientist* (1975).
36. 'Ventricular assist system', *Radioisotope Report*, September 1975, *OECD Nuclear Energy Agency Newsletter*, **6** (1976), 9.
37. 'Artificial heart status report', *OECD Nuclear Energy Agency Newsletter*, **4** (1974), 8.
38. F. Wallner 'Implantable devices and their energy needs', *Second International Symposium on power from radioisotopes*. (1972), Madrid, Spain, p. 763–8.
39. R.E. Avery and J.S. Melville, 'Implantable electrodes for the stimulation of the sciatic nerve', *US Patent* 3,738,868, (June 1973).
40. J.D. Hixon and P. Laurens, 'Design criteria and two year clinical results of Pu-238 fuelled demand pacemaker', *Proceedings of 7th IECEC*, (September 1972), San Diego, California, pp. 765–70.
41. M. Schaldach, R. Thull and R. Blaser, 'Kernenergiebetriebene Batterien für Implantierbare Hergschrittmacker', *Atomikernenergie (ATKE) Bd* **21** (1973), 17–24.
42. 'Environmental draft, praised, criticized', *Nuclear News*, May 1975.
43. 'Nuclear pacemaker market not booming but vendors see stable future', *Nucleonics Week*, 19 February 1976.
44. 'Pacemakers–NRC survey shows doctors dissatisfied', *Nuclear News*, October 1976.
45. 'NRC issues environmental statement on routine use of plutonium powered nuclear pacemakers', *US NRC News Releases* **2**, August 1976.
46. H. Victor, 'The application of radioisotopic power generators in oceanography and marine technology', *Second International Symposium on Power from Radioisotopes*, (1972), Madrid, Spain, p. 53.
47. H. Victor 'Zur Entwicklung von Fortschrittlicken Energieversongungangen für die Meerestechnik', *Meerestechnik*, **1** (1970), 2.
48. H.C. Carney, 'Comparison of strontium-90 and plutonium-238 milliwatt thermoelectric generators', *Second International Symposium on Power from Radioisotopes*, (1972), Madrid, Spain, pp. 143–63.

49. F.E. Rossell, Jr and J.F. Vogt, 'Recent terrestrial and undersea applications of radioisotope thermoelectric generators (RTGs)', *Proceedings of 13th IECEC*, (August 1978), San Diego California, pp. 1969–83.
50. F.E. Rossell, Jr, 'The navy half-watt RTG—a super battery becomes reality', *OECD Nuclear Energy Agency Newsletter*, **6** (1976), 4.
51. V. Merges, 'The development of the radioisotope thermoelectric generator TRISTAN', *Second International Symposium on Power from radioisotopes*, (1972), Madrid, pp. 385–96.
52. J. Gajewski, 'Experimental thermoelectric generator SNAP-3', *Przeglad Elektrotech*, **38**, 252 ff.
53. 'Radioisotope weather station', Radioisotope Report, August 1974, *OECD Nuclear Energy Agency Newsletter*, **75** (1975), 3.
54. Y. Toure, 'Sources d'energie autonomes immergeables', *Entropie*, **52** (1973), 28.
55. E.H. Cooke-Yarborough and F.W. Yeats, 'Efficient thermomechanical generation of electricity from the heat of radioisotopes', *Proceedings of 10th IECEC*, (August 1975), Newark, Delaware, pp. 1003–11.
56. 'RIPPLE at Rathlin O'Birne,' UKAEA-STN/15/74, 13 August 1974.
57. H.T. Fullam, 'Strontium-90 Fluoride Data Sheet', *Battelle Pacific Northwest Laboratories Report* 2284, Fig. 3, October 1977.
58. K.J. Round, 'A review of radioisotope power source development at atomic energy of Canada Ltd', *Second International Symposium on Power from Radioisotopes*, (1972), Madrid, p. 591–612.
59. P.J. Dick, 'Testing of the improved SNAP-19 primary power for advanced space missions', *Second International Symposium on Power from Radioisotopes*, (1972), Madrid, pp. 571–88.
60. G.B. Bradshaw, 'A description of the TRANSIT RTG', *Second International Symposium on Power from Radioisotopes*, (1972), Madrid, p. 545–69.
61. A.W. Fihelly, H.N. Berkow and C.F. Baxter, 'SNAP-19 Nimbus B experience', *Proceedings of 3rd IECEC*, (August 1968), Boulder, Colorado, p. 155.
62. P. Rouklove and V. Truscello, 'Long term tests of SNAP-19 thermoelectric generator', *Proceedings of 7th IECEC*, (September 1972), San Diego, California, pp. 186–93.
63. W.M. Brittain, 'SNAP-19 radioisotope thermoelectric generator thermal/electrical integration with the Viking Lander', *Proceedings 6th IECEC*, (August 1971), Boston, Massachusetts, pp. 1189–99.
64. C.J. Goebel, 'SNAP-19 Pioneer 10 and 11, RTG deep space performance', *Proceedings of 10th IECEC*, (August 1975), Newark, Delaware, pp. 868–72.
65. R.D. Smith, D.R. Seals and K. Farley, 'The Viking Lander power system', *Proceedings of 6th IECEC*, (August 1971), Boston, Massachusetts, pp. 829–36.
66. C.E. Kelly, 'The MHW converter (RTG)', *Proceedings of 10th IECEC*, (August 1975), Newark Delaware, pp. 880–6.
67. R.S. Caputo and V.C. Truscello, 'Long life performance predictions of Si–Ge MHW–RTG', *Proceedings of 8th IECEC*, (August 1973), Philadelphia, pp. 26–38.
68. G. Stapfer, 'The thermoelectric technology program at the Jet Propulsion Laboratory', *Proceedings of Third International Conference on Thermoelectric Energy Conversion*, (March 1980), University of Texas at Arlington, pp. 70–5.
69. P.A. O'Riordan, 'The US Department of Energy's thermoelectric development program for space power', *Proceedings of the Fourth International Conference on Thermoelectric Energy Conversion*, (1982), University of Texas at Arlington, pp. 15–25.
70. P.G. Dix, 'Advances in safety of space nuclear power systems', *Second International Symposium on Power from Radioisotopes*, (1972), Madrid, p. 779–97.
71. J.T. Stadter, 'Safety of MHW', *Proceedings of 10th IECEC*, (August 1975), Newark, Delaware, pp. 887–93.

10

Conclusions

THE continued development and application of thermoelectric generators is assured with the continuation of the US space programme, with RTGs providing onboard power for deep space missions. The lead telluride family of semiconductors has been the mainstay thermoelectric material in space applications. Generators based upon this technology have been used in NIMBUS III, Pioneer, Viking and TRANSIT spacecraft and in the Apollo Lunar Surface Experimental Packages.[1]

The technological difficulties associated with the development of a new thermoelectric technology are illustrated by the 'low-cost high-performance generator'. The project which started in 1973 used gadolinium–selenide thermocouples and generators based upon this advanced technology (Figure 10.1) were intended to provide power for the Jupiter orbiter flight.[2] Originally scheduled for 1981, the mission was delayed for three years due to shuttle launch vehicle problems and serious difficulties encountered in producing stable selenide thermoelectrics have resulted in the selenide project being abandoned.[3] (See section 5.2.2.)

Silicon–germanium alloys have emerged as the most successful thermoelectric material for use in vehicles operating in 'deep space'. Developed at RCA for use in the US multi-hundred watt generators (155 W(e) at an efficiency of 6.5 per cent and a specific power of 3.96 W(e) per kilogram) the first space application of this generator was early in 1976 in US Air Force LES 8/9 communication satellite and followed that year by the Voyager spacecrafts. Silicon–germanium technology has been chosen by the US Department of Energy for use in conjunction with the newly developed 'general purpose heat source (GPHS) to provide on-board power (294 W(e) at an efficiency of 6.7 per cent and a specific power of 5.28 W(e) per kilogram) for the Galileo Jupiter orbiter flight and the International Solar Polar mission. The Department of Energy's goal for the late 1980s of a specific power of 7.7 W(e) per kilogram appears to be met by the new Fairchild modular RTG design.

It seems certain that in marine applications the requirement for isotope powered generators in submerged situations will increase principally due to their military application, while their prospect in surface and terrestrial applications is not so certain. Although there are no justified grounds for environmental anxiety, any proliferation in applications that might be expected to accompany a breakthrough in material performance is likely to be met by a strong lobby against large-scale deployment of radioactive sources. From an economic point of view, it seems unlikely that radioisotope generators

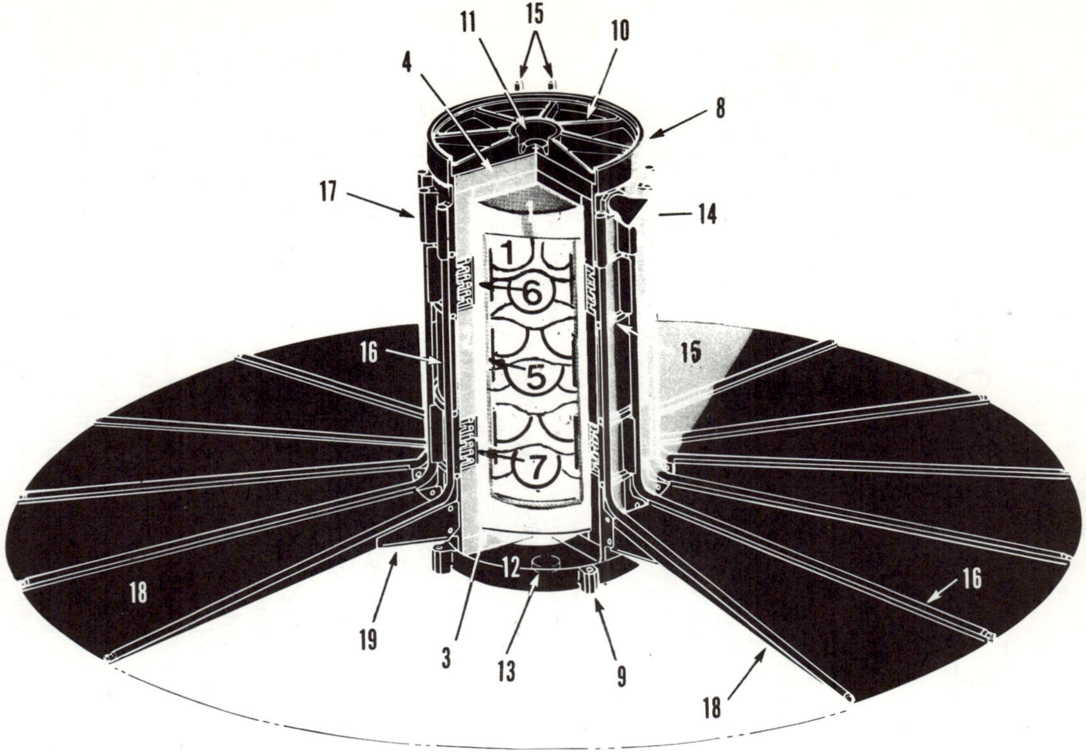

Figure 10.1 The selenide isotope generator 'SIG'. (Courtesy Teledyne Energy Systems, USA.)

1. Radioisotope heat source (MHW)
2. Side thermal insulation
3. High temperature load bearing end thermal insulation
4. Low temperature load bearing end thermal insulation
5. Insulation support sleeve
6. T/E module support frame
8. Housing shell structure
9. Mounting lugs
10. End cover with system vent
11. System vent device
12. End cover with electrical power output connector
13. Electrical power output connector
14. Outgassing pinch-off tube
15. Auxiliary cooling system circuit
16. Heat pipes
17. Heat pipe reservoirs
18. Fin structure

can be justified in situations where air is available to burn hydrocarbon fuel. In situations where it is difficult to get fuel to the site, a thermomechanical generator presents a possible alternative to thermoelectric generators.

In medical application, particularly the nuclear-powered cardiac pacemaker, the future is more promising. Although some physicians have expressed dissenting views over the need for such a device, in the UK the DHSS has supported a programme of research into a nuclear-powered battery-operated cardiac pacemaker. An important factor is the reduced risk to the patient resulting from the use of nuclear-powered pacemaker rather than chemical batteries. As for environmental objections to the wide-scale use of plutonium-powered cardiac pacemakers, the US Atomic Energy Commission Directorate of Licensing has made it clear that their social and economic benefits far outweigh the social and economic costs.

The present restricted commercial applications of both thermoelectric generators and refrigerators can be largely attributed to their relatively poor performance compared to conventional electric power generators and vapour compression cooling devices. Existing thermoelectric generators with conversion efficiencies of ~7 per cent and thermoelectric refrigerators with a coefficient of performance of less than about 0.5 are just not competitive in the majority of applications. In order for a thermoelectric generator to attain a competitive status it must operate with an efficiency of ~20 per cent; this would require a material ZT value in a device which operates over the range 300 to 1000 K of about 2. This is double current material value.

The development of thermoelectric refrigeration has concentrated on device technology rather than on improving the performance of thermoelectric materials, and today's materials, apart from minor modifications, are essentially the same as those employed in thermoelectric coolers in the early 1960s. Materials based upon the bismuth telluride alloys have been used almost exclusively in the fabrication of thermocouples for refrigeration. *N*- and *p*-type bismuth telluride have room temperature average figures of merit of about $3 \times 10^{-3} K^{-1}$, while values at least two or three times higher are needed to make thermoelectric cooling systems competitive with those based upon vapour compression. However, a wide range of thermoelectric cooling modules is available, some of which can operate at hot-side temperatures which are outside the range of compression systems. Technological problems associated with establishing and maintaining good thermal contact between heat transfer surfaces of the cooling systems persist. Nevertheless, cooling devices such as the Borg–Warner air conditioner are enjoying an encouraging level of commercial success.

The most optimistic value of predicted improvement in material performance falls short of 'competitive' values. Nevertheless, the recent reported improvement in the Z values of Si–Ge alloys which accompany the addition of groups III–V elements and the use of small grain size material should lead to a significant increase in thermoelectric generator performance.

As to the possible large scale generation of thermoelectric power, it appears very unlikely within the foreseeable future. However, as the price of fossil fuels increases and conventional methods of producing electrical power become more costly, unconventional methods such as thermoelectric generation become progressively more attractive.

REFERENCES

1. P. O'Riordan, 'Department of Energy Executive Summary', *Proceedings of 13th IECEC*, (1978), San Diego, California, pp. 1525–26.
2. G. Stapfer, 'The thermoelectric technology programme at the Jet Propulsion Laboratory', *Proceedings of the Third International Conference on Thermoelectric Energy Conversion*, (March 1980), University of Texas at Arlington, pp. 70–5.
3. A.R. Lieberman, project engineer, Teledyne Energy Systems, Timonium, Maryland. (1981). Private communication.
4. P.A. O'Riordan, 'The US Department of Energy's thermoelectric development program for space power', *Proceedings of the Fourth International Conference on Thermoelectric Energy Conversion*, (1982), University of Texas at Arlington, pp. 15–25.

Appendix—Symbols

B	magnetic field
C	phonon heat capacity per unit volume
C_p	specific heat at constant pressure
C'_p	heat capacity per unit volume at constant pressure
D_a, D_b	density of states in accepter and donor levels
E	energy of charge carriers (electrons or holes)
E_F	Fermi energy
E_g	energy band gap
ΔE_a, ΔE_b	energy corresponding to accepter and donor levels
F	form factor
$F_l(\xi)$	Fermi integrals
$\vec{G}$	reciprocal lattice vector
I	electric current
L	Lorenz number, Casimir length, grain size
$\mathscr{L}$	Lorenz factor
N	concentration of electrons in the valence band
N_d	concentration of electrons in the impurity levels
N_e, N_h	number of extrema in conduction or valence band
N_v	number of valleys
Q	quantity of heat transfer
$\dot{Q}$	rate of heat transfer
R	electrical resistance
R_H	Hall coefficient
R_L	load resistance
T	absolute temperature
$\bar{T}$	mean temperature
ΔT	temperature difference
ΔT_{max}	maximum temperature difference
V	voltage
V_H	Hall voltage
Z	thermoelectric figure of merit of an individual material
Z_c	thermoelectric figure of merit for a junction or couple formed from two materials
ZT	dimensionless figure of merit
a	mass anisotropy factor
b	ratio of strength of ionized impurity scattering to lattice scattering, sample thickness
c_{ll}	elastic constant

e	electronic charge
$_e$(subscript)	n-type, electron or negative
f_i	fraction of atoms of mass m_i
h	Planck constant
$_h$(subscript)	p-type, hole or positive
$\hbar$	$h/2\pi$
k	Boltzmann constant
k_0	ratio of phonon relaxation time for Umklapp and normal processes
l	phonon mean free path
m	mass of electron, mass of atom, area to length ratio of thermocouple arm
m^*	density of states effective mass
m_e^*, m_h^*	electron and hole density of states effective masses
m_c, m_i	conductivity or inertial effective mass
m_s	density of states mass for a single valley
m_t	overall density of states mass
n	carrier concentration, number of thermocouples
$\vec{q}$	phonon wave vector
r	scattering parameter defined by $\Delta(E)=\text{constant } E^r$, Δ being the carrier mean free path
r_H	Hall factor
s	carrier scattering parameter defined by $\tau(E)=\text{constant } E^s$
v_s	phonon wave velocity
α	Seebeck or thermoelectric coefficient of an individual material
α'	reduced Seebeck coefficient, thermal diffusivity
α_{ab}, α_c	Seebeck coefficient of a junction ab or couple c formed from two materials a and b
α_e	Seebeck coefficient due to electrons
α_h	Seebeck coefficient due to holes
β	Thomson coefficient, material parameter
Γ	disorder parameter
δ	Dirac delta function, combination of Fermi integrals
δ_1	wave amplitude ratio
ε	coefficient of performance of a refrigerator
ε_{max}	maximum coefficient of performance
$\varepsilon_e, \varepsilon_h, \varepsilon_1$	deformation potential constants
η	reduced carrier energy
η_c	Carnot efficiency
Θ	phonon characteristic temperature
λ	thermal conductivity
λ'	thermal conductance
λ_L	lattice component of thermal conductivity
λ_0	thermal conductivity of notional, perfect, large single crystal
λ_p	polar component; electron (λ_e) or hole (λ_h) of thermal conductivity
λ_b	bipolar component of thermal conductivity
μ	carrier mobility
μ_c	mobility in classical (low concentration) limit
μ_e, μ_h	electron and hole mobility

μ_H	Hall mobility
ξ	reduced Fermi energy
ξ_g	reduced energy gap
ξ_{opt}	optimum value of reduced Fermi energy
π	Peltier coefficient of an individual material
π_{ab}	Peltier coefficient of a junction from materials *a* and *b*
ρ	electrical resistivity
σ	electrical conductivity
σ'	reduced electrical conductivity
σ_e	electrical conductivity due to electrons
σ_h	electrical conductivity due to holes
τ	phonon relaxation time, charge carrier relaxation time
τ_{ac}	carrier—acoustic phonon scattering relaxation time
τ_{dis}	phonon—disorder scattering relaxation time
τ_{ep}	phonon—electron (hole) scattering relaxation time
τ_{imp}	carrier—impurity scattering relaxation time
τ_N	phonon relaxation time for normal processes
τ_U	phonon relaxation time for *U*-processes
ϕ	generation efficiency
ϕ_p	generator efficiency at maximum power output
ϕ_s	system efficiency of a thermoelectric device
ϕ_{max}	maximum generator efficiency
Φ	phase difference
χ_m	concentration of minority component
Ω_0	volume per atom
ω	phonon frequency

Index